INTRODUCTION TO MOLECULAR DYNAMICS AND CHEMICAL KINETICS

Phys

INTRODUCTION TO MOLECULAR DYNAMICS AND CHEMICAL KINETICS

GERT D. BILLING
Department of Chemistry
H. C. Ørsted Institute
University of Copenhagen

KURT V. MIKKELSEN
Department of Chemistry
Århus University

A Wiley-Interscience Publication

JOHN WILEY & SONS, INC.

New York • Chichester • Brisbane • Toronto • Singapore

This text is printed on acid-free paper.

Copyright © 1996 by John Wiley & Sons, Inc.

All rights reserved. Published simultaneously in Canada.

Library of Congress Cataloging in Publication Data:
Billing, Gert D.
    Introduction to molecular dynamics and chemical kinetics / Gert D.
Billing and Kurt V. Mikkelsen.
        p.   cm.
    ISBN 0-471-12739-6 (alk. paper)
    1. Molecular dynamics.   2. Chemical kinetics.   I. Mikkelsen, Kurt
V.   II. Title.
QD461.B573 1996
541.3′94—dc20   95-23950
                                                          CIP

Printed in the United States of America

10 9 8 7 6 5 4 3

To Inge-Lise and Maryanne

CONTENTS

PREFACE

Most standard textbooks on physical chemistry contain a section on molecular reaction dynamics. However, they hardly go beyond the introduction of a collision cross section and a discussion of reaction mechanisms and rate constants through the Arrhenius rate law. Quantitative calculations of rates are usually discussed only in the context of transition state theory. The present monograph also covers these areas, but in addition includes concepts from molecular dynamics calculations as well as methods for treating reactions in solution.

The monograph originates from our efforts to establish an introductory course on molecular reaction dynamics encompassing reactions in the gas phase, on surfaces, and in solution. We were not able to find either a textbook or other suitable material that could cover the topics we considered to be important. Therefore, we began writing notes on the different topics; these notes have led to this monograph.

The course has been given to undergraduate (third year) students at the Universities of Copenhagen and Århus. It is assumed that the student has some basic knowledge of mathematics, electrostatic theory, classical mechanics, and statistical mechanics. However, we have in the appendices given a brief overview of many of the necessary tools.

We propose a system of molecular units, which are convenient for molecular dynamics calculations. However, since we realize that most students are accustomed to other standards, physical constants are also given in more conventional units. Many of the details involved in the derivations in the text are given in the form of exercises. Other exercises are given to aid general understanding of the effects under discussion. Therefore, it is highly recommended to work through these. The "Answers" section gives fairly detailed explanations on how to solve the exercises.

Essentially all the necessary mathematical derivations for obtaining the formulas are given. It is our belief that most students are interested in seeing how final expressions are derived, thereby enabling them to judge the underlying approximations. The first ten chapters cover basic concepts and approaches mainly related to models for chemical reactions in the gas phase. The following six chapters comprise models for chemical reactions on surfaces and in solution.

Chapter 1 is an introduction. Chapter 2 covers interaction potentials commonly used in molecular reaction dynamics. The following two chapters describe the concept of relative motion and the collisional approach for a chemical reaction in the gas phase. Chapter 5, which should be studied along with Appendix D on statistical mechanics, considers the necessary ingredients (partition functions) for the next three chapters. Chapters 6 and 7 include the presentation of transition state theory, either in its simple form or its generalized counterpart. Unimolecular reactions are discussed in Chapter 8. Chapter 9 shows the basic building blocks involved in performing molecular reactions calculations. Nonadiabatic transitions are discussed in Chapter 10, leading to Chapter 11 concerning surface kinetics. The chapters covering chemical reactions in solution start out by presenting the motion of particles in solution; Appendix C on Laplace transform provides some of the basic mathematical tools. Chapter 13 studies the energetic changes involved when solvating a molecule; here one should confer with Appendix E, "Notes on Solvent Model." Chapter 14 studies the use of transition state theory in solution along with models for diffusion-influenced reactions. Chapters 15 and 16 deal with two important areas: (1) Kramers' theory for the effect of the viscosity of the solvent on chemical reactions and (2) Marcus' theory for electron transfer in solution. Appendix F, "Electrostatic Energy of a Polarized Dielectric," covers some of the underlying electrostatic terms and definitions needed for understanding Marcus' theory. Maryanne Kmit's review of the manuscript is gratefully acknowledged. The same goes for all the students, who, during the courses we have held, have given suggestions and found misprints in the notes.

GERT D. BILLING
KURT V. MIKKELSEN

Copenhagen and Århus, 1995

1

INTRODUCTION

The ultimate goal of a molecular dynamical approach to chemical reaction dynamics is to be able to understand and calculate the rate of chemical reactions from first principles, i.e., given a specific interaction potential for the nuclear motion one should in principle be able to obtain the probabilities, cross sections, and rate constants for fundamental elementary reaction processes by solving the equations of motion for the system. This goal has so far only been realized for very few reactions, the reason being that the complete determination of the electronically adiabatic Born-Oppenheimer surface is computationally an immense task. Furthermore the solution to the Schrödinger equation for the nuclear motion has also turned out to be quite a formidable undertaking. Therefore much of the understanding of chemical reactivity at a molecular level has come about by using model potential energy surfaces, approximate dynamical methods (e.g., classical mechanics) or various statistical assumptions and models, such as those introduced in the transition state theory and theories for unimolecular reactions. This research area is today a very active one, where the experimental and theoretical techniques for probing systems at a molecular and/or time-resolved level are becoming more and more refined.

For simple three-center gas phase reactions, it is possible to construct reliable semiempirical potential energy surfaces including both experimental and ab initio data when available. With the full multidimensional potential energy surface (PES) classical trajectory computations are often accurate for the total reaction rate, but less reliable for state-to-state cross sections. Thus classical trajectory studies have become an important standard theoretical tool for converting the information contained in the potential energy surface to measurable quantities. For larger systems, where less information is available, it is possible

1

to use transition state concepts. Here only information on geometries and frequencies at the transition state is needed. For reactions in solution, even less is known and hence the dynamical approach that can be used is based upon theories such as the dielectric response and Brownian motion. Other dynamical issues of interest for chemical reactivity in solution are the influence of diffusion processes, frictional effects, and dielectric relaxation.

2

INTERACTION POTENTIALS

The outcome of any chemical reaction is determined by physical properties such as mass, temperature, and the "chemical" property known as the interaction potential. The potential determines the forces acting on the various atoms and decides, in turn, whether it is possible with a given energy content in the system to break one bond and form a new one, i.e., whether a chemical reaction can take place.

2.1 THE HARMONIC POTENTIAL

The simplest potential for the interaction of two atoms in a diatomic molecule is obtained by using Hooke's law, in which the restoring force is simply proportional to the displacement of the intermolecular distance from equilibrium, i.e.,

$$-F(r) = \frac{\partial V(r)}{\partial r} = k(r - r_e) \tag{2.1}$$

where k is the force constant and r_e the equilibrium distance. Thus this force law gives a potential of the type:

$$V(r) = \tfrac{1}{2}k(r - r_e)^2 \tag{2.2}$$

This potential is denoted "harmonic." It represents the interaction between the

3

two atoms in the vicinity of the equilibrium distance r_e, but not, of course, for those at large asymptotic distances. It is a model potential that when used in the nuclear Schrödinger equation

$$-\frac{\hbar^2}{2\mu}\frac{\partial^2}{\partial r^2}\psi + V(r)\psi = E\psi \tag{2.3}$$

gives the energy levels in terms of the vibrational quantum number v as:

$$E_v = \hbar\omega_e(v + \tfrac{1}{2}) \tag{2.4}$$

where $v = 0, 1, 2, \ldots$. The reduced mass μ of the diatomic molecule is

$$\mu = \frac{m_1 m_2}{m_1 + m_2} \tag{2.5}$$

The vibrational wavenumber ω_e is related to the force constant and the reduced mass by the equation

$$\omega_e = \sqrt{\frac{k}{\mu}} \tag{2.6}$$

Since information about the frequency $\omega_e = 2\pi\nu_e$ is available from spectroscopic measurements we see that the force constant and hence the potential in this approximation can be obtained experimentally. However, as mentioned, the harmonic potential only represents the interaction well in a small region of r-space or equivalently the spectra are only approximately reproduced by Eq. 2.4. In order to extend the range we may add higher order potential terms, i.e., use

$$V(r) = \tfrac{1}{2}k(r - r_e)^2 + f_3(r - r_e)^3 + f_4(r - r_e)^4 \tag{2.7}$$

This expression includes cubic and higher order terms in the displacement coordinate, and the higher order force constants f_3 and f_4 are expected to be smaller than the harmonic $f_2 = k/2$. If we include these higher order terms in the solution to the Schrödinger equation, we obtain the following modified expression for the energy levels:

$$E_v = \hbar\omega_e\left[(v + \tfrac{1}{2}) - x_e(v + \tfrac{1}{2})^2 + y_e(v + \tfrac{1}{2})^3\right] \tag{2.8}$$

Although the anharmonic corrections x_e and y_e in Table 2.1 may look small, the

TABLE 2.1. Harmonic Frequencies and Anharmonic Corrections for Some Diatomic Molecules

Molecule	ω_e (cm^{-1})	x_e	y_e	r_e (Å)	D_e (eV)
H_2	4395.24	0.02685	6.60×10^{-5}	0.7417	4.7477[a]
N_2	2359.61	0.00613	3.18×10^{-6}	1.094	9.905
HCl	2989.74	0.01741	1.87×10^{-5}	1.275	4.618
HF	4138.52	0.02176	2.37×10^{-4}	0.9171	6.1223[b]
O_2	1580.36	0.00764	3.45×10^{-5}	1.207	5.2136
CO	2170.21	0.00620	1.42×10^{-5}	1.128	11.224

Data taken from refs. [1–4].
[a]From ref. [3].
[b]From ref. [4].

terms become substantial as the vibrational quantum number increases. Hence the expansion in the power series Eq. (2.7) cannot be used for $r \rightarrow \infty$. To describe this limit more correctly one has to look for an analytical expression for which the Schrödinger equation is solvable in terms of known functions. Such an empirical potential, which qualitatively has the correct asymptotic behavior, is the Morse potential [5] (see Fig. 2.1):

$$V(r) = D_e\{1 - \exp[-\beta(r - r_e)]\}^2 \tag{2.9}$$

where D_e and β are parameters. The parameter D_e is the dissociation energy D_0 plus the zero point energy E_0. The Morse parameter β is related to the harmonic force constant by:

$$k = 2D_e\beta^2 \tag{2.10}$$

Although the Morse potential is not able to give an altogether satisfactory fit to the actual interaction potential for a large range of r-values it is an extremely useful approximate potential, which with an appropriate choice of parameters can be made to fit the interaction in desired r-ranges, e.g., for large r-values or for values near r_e. The Morse potential describes the chemical binding of the two atoms and the dissociation properly. Repulsive antibonding potential curves can be represented by an anti-Morse function of the type

$$V(r) = \tfrac{1}{2}D_e(1 + \exp(-\beta(r - r_e)))^2 \tag{2.11}$$

Several other empirical potential energy functions have been suggested (see for example ref. 2), one of which is the modified Rydberg potential:

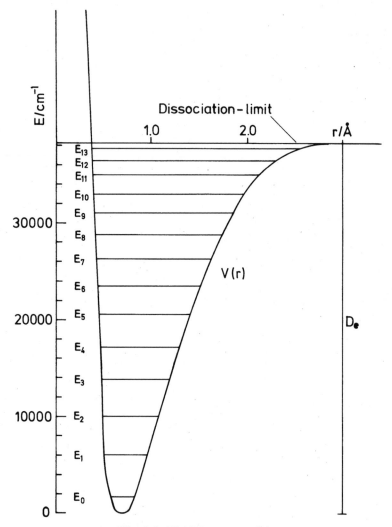

Fig. 2.1. The Morse potential.

$$V(r) = -D_e(1 + \beta(r - r_e)) \exp(-\beta(r - r_e)), \qquad (2.12)$$

where $\beta = \sqrt{k/2D_e}$. The fit that can be obtained with this potential is usually very good and can easily be improved by adding additional parameters

$$V(r) = -D_e[1 + a_1(r - r_e) + a_2(r - r_e)^2 + \ldots] \exp(-\gamma(r - r_e)]. \qquad (2.13)$$

These potentials will, like the Morse potential, be able to describe bond breaking/formation.

2.2 THREE-BODY INTERACTION

In order for a process to be called a chemical reaction at least three atoms must be involved, so that one chemical bond can be formed when another is broken. One must generalize the two-atom concept to a polyatomic one and then in order to obtain rate constants, investigate the role of energy content, initial quantum numbers, steric factors, and so on, one must learn how to compute classical trajectories or give approximate quantum methods, which utilize the information given in the potential. In this manner it is possible to obtain information on chemical reactivity from "first principles."

How can one generalize the concept of an interatomic potential? One needs to know the interaction energy in the full configuration space; for N atoms this is of the dimension $3N-6$. Thus for three atoms (AB and C) the potential energy surface will be a function of three coordinates, for instance the three distances R_{AB}, R_{AC}, and R_{BC}. In order to visualize the potential energy surface one usually considers the collinear approach, for which the potential energy surface can be considered to be a function of only two distances (the third being trivially given as the sum of the other two).

Such a potential plot is shown in Fig. 2.2. We see that the surface changes smoothly from reactants A + BC to products AB + C. The surface is characterized by a minimum energy path, a saddle point, and asymptotic correct behavior, i.e., as the diatomic BC and AB potentials. The potential contour maps are often shown using the "kinematic" skewing angle ξ, where

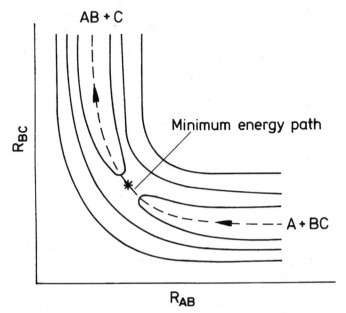

Fig. 2.2. Potential contour plot for a collinear A + BC collision.

$$\tan(\xi) = \sqrt{\frac{m_B(m_A + m_B + m_C)}{m_A m_C}} \qquad (2.14)$$

rather than the 90° angle used in Fig. 2.2.

The energy profile along the minimum energy path is shown in Fig. 2.3. The chemical reaction for a reaction of the type:

$$D + H_2 \rightarrow DH + H \qquad (2.15)$$

takes place not in two steps (breaking one bond and forming a new) but rather by a simultaneous process where one bond is gradually formed while the other is broken (so-called concerted motion of the atoms). Thus the energy barrier—here defined as the maximum along the minimum energy path—for forming the AB molecule is much smaller than the dissociation energy of the reactant molecule. For the simplest reactive systems (i.e., systems containing few electrons) the potential energy surface is in principle available from high level ab initio electronic structure calculations. For such calculations to be chemically accurate, and to predict, e.g., the energy barrier with an accuracy better than 1 kcal/mol, they must include electronic correlation and are therefore still rather elaborate calculations.

Another aspect that makes these calculations time consuming and therefore expensive is the fact that the complete potential energy surface must be scanned. Thus for 3 atoms and just 10 values for each distance parameter we would need to perform the calculation at about 1000 points (sets of distances). After having completed the calculations one would have to fit an analytical expression to the points—an expression that should describe the three asymptotic channels correctly (a channel is defined as an outcome of a reaction). Such calculations have so far been completed for very few systems. The best investigated systems are H_3 and FH_2. Thus even today, where high-speed computers have been available for more than a decade, most calculations of chemical reactivity are carried out on semiempirical surfaces, in which both experimental (i.e., spectroscopic information) and ab initio points, when available, are included.

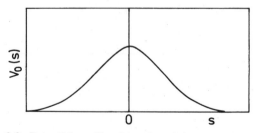

Fig. 2.3. Potential profile along the minimum energy path.

Hence model potentials are needed, which interpolates correctly between the asymptotic regions and that include a saddle point region (an activation barrier). Such a model potential could then be used as the zeroth order approximation to the "true" potential and corrected accordingly when more information became available. For a collinear reaction one could imagine constructing a potential such as

$$V(R_{BC}, R_{AB}) = V_1(R_{BC})f_1(R_{AB}) + V_2(R_{AB})f_2(R_{BC}) \qquad (2.16)$$

where V_1 and V_2 are for example Morse potentials and f_i $(i = 1, 2)$ are switching functions for which f_1 is zero for values of large R_{BC}, and unity for values of large R_{AB}. This form will correctly describe the asymptotic behavior of the potential. The switching functions will have to be estimated and fitted to experimental information concerning the saddle point energy. Note that the diatomic asymptotic potentials play an important role in this semiempirical potential. Another more sophisticated potential that has the same property is the LEPS (London-Eyring-Polanyi-Sato) [6] potential, which was derived for a three-electron S-state atomic system. This potential represents the asymptotic diatomic potentials in terms of a coulombic (Q) and an exchange (J) contribution, i.e.,

$$V(R_{AB}) = Q_{AB} \pm J_{AB}, \qquad (2.17)$$

where the plus sign refers to the singlet electronic ground state and the minus sign to the repulsive triplet state (note that J is negative); see Fig. 2.4. The LEPS potential is now:

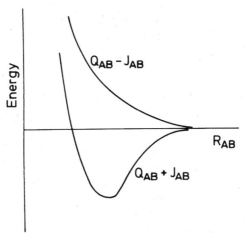

Fig. 2.4. Singlet and triplet states represented by Morse and anti-Morse potential curves.

$$V_{LEPS} = Q_{AB} + Q_{BC} + Q_{AC}$$
$$\pm \sqrt{\tfrac{1}{2}(J_{AB} - J_{BC})^2 + \tfrac{1}{2}(J_{BC} - J_{CA})^2 + \tfrac{1}{2}(J_{CA} - J_{AB})^2} \quad (2.18)$$

When one of the atoms is positioned at infinity, the coulomb and exchange terms for the two bonds involving this atom vanish and we are left with expression (2.17). Thus expression (2.18) behaves as it should asymptotically. By introducing the Sato [7] parameters S_{AB} such that:

$$V(R_{AB}) = \frac{1}{1 \pm S_{AB}}(Q_{AB} \pm J_{AB}), \quad (2.19)$$

we obtain additional flexibility in the expression and are then able to fit the saddle point region better. Since the LEPS potential has the correct global behavior it is a good starting point for an accurate representation of many three-body interaction potentials. Correction terms can then be added if more flexibility is needed (see for example ref. [3]). Approximating the asymptotic ground state by a Morse function and the excited state by an anti-Morse function we obtain:

$$Q_{AB} + J_{AB} = D_e\{1 - \exp[-\beta(R_{AB} - R_{AB}^e)]\}^2 - D_e \quad (2.20)$$

$$Q_{AB} - J_{AB} = \tfrac{1}{2}D_e\{1 + \exp[-\beta(R_{AB} - R_{AB}^e)]\}^2 - \tfrac{1}{2}D_e \quad (2.21)$$

i.e., we obtain the coulomb and exchange terms from the spectroscopically determined diatomic interaction potentials. Hence the triatomic potential energy surface is, in this model, uniquely determined. In order to estimate the quality of a given potential energy surface, however, one needs to solve the dynamical problem, i.e., to obtain quantitites such as cross sections, reaction probabilitites, and rate constants, which can be compared with experimental data. The dynamical problem can be solved within various approximations, such as for instance within a transiton state approximation or using approximate classical dynamics. The first method allows for an analytical expression for the rate constant, whereas the latter requires numerical, but straightforward, trajectory calculations. From the potential contour maps and the location of the reaction barrier it is, however, possible to explain qualitatively why vibrational excitation in some cases enhances the reaction whereas in other cases the reaction probability is increased by increasing translational energy. Fig. 2.5 shows that the former is the case of a so-called late barrier along the reaction path, and the latter that for which the barrier is early, i.e., located in the entrance channel.

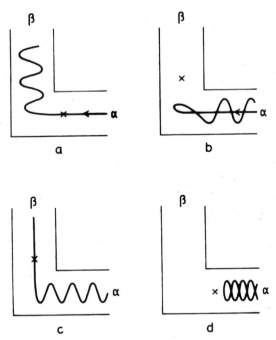

Fig. 2.5. Effect of the initial excitation on the reactivity for early and late barriers. The position of the barrier is indicated by (x) and the trajectory starts in channel α. Vibrational excitation favors the reaction if the barrier is late along the reaction path (**c**) and translational energy favors the reaction if the barrier is early (**a**). These are the celebrated Polanyi rules [8].

REFERENCES

[1] G. Herzberg, "Spectra of Diatomic Molecules," Van Nostrand, Princeton, NJ, 1950; K. P. Huber and G. Herzberg, "Molecular Spectra and Molecular Structure, IV. Constants of Diatomic Molecules," Van Nostrand, New York, 1979.

[2] J. N. Murrell, S. Carter, S. C. Farantos, P. Huxley and A. J. C. Varandas, "Molecular Potential Energy Functions," John Wiley & Sons, London, 1984.

[3] B. Liu, J. Chem. Phys. **58,**1925(1973); P. Siegbahn and B. Liu, J. Chem. Phys. **68,**2457(1978); D. G. Truhlar and C. J. Horowitz, J. Chem. Phys. **68,**2466(1978).

[4] G. C. Lynch, R. Steckler, D. W. Schwenke, A. J. C. Varandas, D. G. Truhlar and B. C. Garrett, J. Chem. Phys. **94,**7136(1991).

[5] P. M. Morse, Phys. Rev. **34,**57(1929); H. S. Heaps and G. Herzberg, Z. Physik. **133,**48(1952).

[6] H. Eyring and M. Polanyi, Z. Physik. Chem. **B12,**279(1931); F. London, Z. Elektrochem. **35,**552(1929).

[7] S. Sato, J. Chem. Phys. **23,**592, 2465(1955).

[8] J. C. Polanyi, Accts. Chem. Res. **5,**161(1972).

SUGGESTED READING

D. M. Hirst, "Potential Energy Surfaces: Molecular Structure and Reaction Dynamics," Taylor & Francis, London, 1985.

R. D. Levine and R. B. Bernstein, "Molecular Reaction Dynamics and Chemical Reactivity," Oxford University Press, Oxford 1987.

EXERCISES

1. Calculate the dissociation energy and the Morse parameter β from the spectroscopic constants ω_e and x_e in Table 2.1. Compare the result with the value given for D_e in Table 2.1. Discuss differences. (Use $x_e = \hbar\omega_e/4D_e$).

2. Derive the expression (2.14) for the angle ξ. Hint: Introduce the kinetic energy (T) for collinear A + BC such that $T = \tilde{\mu}/2(\dot{x}^2 + \dot{y}^2)$. Indicate the distances R_{BC} and cR_{AB} in an xy coordinate system. Determine the scaling parameter c.

3. Explain using Fig. 2.5 the physical reason for the Polanyi-rules. Consult eventually ref. [8].

3

RELATIVE MOTION

Chemical reaction dynamics deals with the theoretical treatment of the collision of two systems (atoms or molecules). In order to simplify the treatment, and thus the notation, it is convenient to deal with relative motion, i.e., the motion of one atom, the collider, relative to that of another atom, the target. Furthermore, the concepts can readily be carried over to molecules, with the modification that we consider instead the motion of the center of mass of the molecules. Thus the two atoms and/or molecules under consideration have the masses m_1 and m_2 and their position and velocity vectors are $\mathbf{r}_i$ and $\mathbf{v}_i$ ($i = 1, 2$).

The collision problem can now be solved and simplified by introducing the two conservation constraints, i.e., for a system without external forces the conservation of total momentum and energy of the system. Thus introducing the total momentum as $\mathbf{p}$ we have:

$$\mathbf{p} = m_1\mathbf{v}_1 + m_2\mathbf{v}_2 \tag{3.1}$$

Thus the velocities are not independent but must change so that the above constraint is fulfilled. We define the relative velocity

$$\mathbf{v} = \mathbf{v}_1 - \mathbf{v}_2 \tag{3.2}$$

These two equations can be used to express the velocities $\mathbf{v}_i$ through $\mathbf{p}$ and $\mathbf{v}$, i.e.,

$$\mathbf{v}_1 = \frac{\mathbf{p}}{m_1 + m_2} + \frac{\mu}{m_1}\mathbf{v} \tag{3.3}$$

$$\mathbf{v}_2 = \frac{\mathbf{p}}{m_1 + m_2} - \frac{\mu}{m_2}\mathbf{v} \tag{3.4}$$

where μ is the reduced mass

$$\mu = \frac{m_1 m_2}{m_1 + m_2} \tag{3.5}$$

Considering the kinetic energy, we obtain

$$E_{\mathrm{kin}} = \tfrac{1}{2}m_1\mathbf{v}_1^2 + \tfrac{1}{2}m_2\mathbf{v}_2^2 = \tfrac{1}{2}(m_1 + m_2)\,\mathbf{c}^2 + \tfrac{1}{2}\mu\mathbf{v}^2 \tag{3.6}$$

where $\mathbf{c} = \mathbf{p}/(m_1 + m_2)$ is the velocity of the entire system. The last term in Eq. (3.6) is the kinetic energy of the relative motion. If we consider an elastic collision the kinetic energy before and after the collision is the same. Since the momentum $\mathbf{p}$ is also conserved we see that this collision will also conserve the relative kinetic energy, i.e.,

$$|\mathbf{v}'| = |\mathbf{v}| \tag{3.7}$$

where $\mathbf{v}'$ denotes the relative velocity after the collision. Thus only the direction of the relative velocity changes during the collision. This is illustrated in Fig. 3.1.

Introducing now the center of mass position vector $\mathbf{R}_c$ and the relative position vector $\mathbf{R}$ as

$$\mathbf{R}_c = \frac{m_1\mathbf{r}_1 + m_2\mathbf{r}_2}{m_1 + m_2} \tag{3.8}$$

$$\mathbf{R} = \mathbf{r}_1 - \mathbf{r}_2 \tag{3.9}$$

we see that the first term in Eq. (3.6) is the kinetic energy of the center of mass

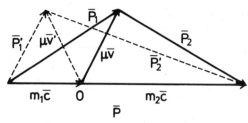

Fig. 3.1. Position of the relative momentum vectors before and after a collision.

motion and the last term is, as mentioned above, that of the relative motion. If the force exerted on atom (molecule) 1 by atom (molecule) 2 is denoted by $F_{12}(R)$ we find from Eq. (3.3)

$$\frac{d\mathbf{p}_1}{dt} = m_1 \frac{d\mathbf{v}_1}{dt} = F_{12}(R) = \mu \frac{d^2\mathbf{R}}{dt^2} \tag{3.10}$$

Thus the relative motion of atom (molecule) 1 with respect to atom (molecule) 2 (given by $\mathbf{R}(t)$) is as if it had a mass μ and was acted on by a force F_{12}. This observation can be used to convert the two-particle problem into a one-particle problem. Thus with respect to particle 2 the scattering process appears as shown in Fig. 3.2. This means we can consider the situation to be one where particle 2 is at rest and particle 1 has the reduced mass μ.

The distance b (the impact parameter) is (see Fig. 3.2) the perpendicular distance from particle 2 to the line generated by the velocity vector $\mathbf{v}$. During the collision the magnitudes of $R(t)$ and the angle $\psi(t)$ (see Fig. 3.3) change and the kinetic energy can be written as

$$E_{\text{kin}} = \tfrac{1}{2}\mu\dot{\mathbf{R}}^2 = \tfrac{1}{2}\mu(\dot{R}^2 + R^2\dot{\psi}^2) \tag{3.11}$$

where $\dot{R} = (dR/dt)$ and $\dot{\psi} = (d\psi/dt)$. Thus the kinetic energy is a sum of the radial kinetic energy (first term) and the centrifugal kinetic energy (second term). Introducing now

$$\dot{\psi} = \frac{b|\mathbf{v}|}{R^2} \tag{3.12}$$

we obtain

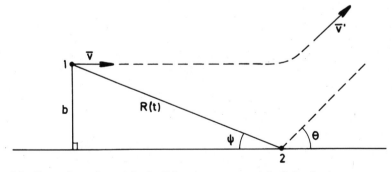

Fig. 3.2. Scattering of particle 1 with respect to particle 2 in the impact parameter picture. The incident (initial) velocity is $v = |\mathbf{v}|$ and the final velocity $v' = |\mathbf{v}'|$.

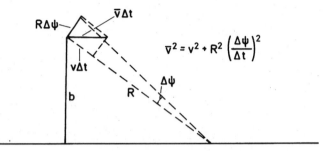

Fig. 3.3. The relative motion is divided into motion along the center of mass distance and orbital motion perpendicular to the center of mass motion.

$$E_{kin} = \frac{1}{2}\mu\dot{R}^2 + \frac{1}{2}\mu|\mathbf{v}|^2\frac{b^2}{R^2} \tag{3.13}$$

Introducing the total energy

$$E = E_{kin} + V(R), \tag{3.14}$$

and the orbital angular momentum $L = \mu b|\mathbf{v}|$ we obtain

$$E = \frac{1}{2}\mu\dot{R}^2 + \frac{L^2}{2\mu R^2} + V(R) \tag{3.15}$$

The last two terms can be considered to be an effective potential energy

$$V_{eff}(R) = V(R) + \frac{L^2}{2\mu R^2} \tag{3.16}$$

As particle 1 approaches the target we see that part of the kinetic energy is converted into orbital energy and for large values of the angular momentum L the centrifugal term builds up a centrifugal barrier that prevents the two particles from coming close to each other (see Fig. 3.4). The approach we have just considered is a classical mechanical treatment of the motion. In a quantum mechanical treatment the angular momentum would be given as

$$L = \hbar\sqrt{l(l+1)} \tag{3.17}$$

where l takes on the values $0, 1, 2 \dots$. Note that the treatment of the orbital angular motion is similar to the treatment of rotational motion of a diatomic molecule but with j (the rotational angular momentum) replaced by the angular momentum quantum number l.

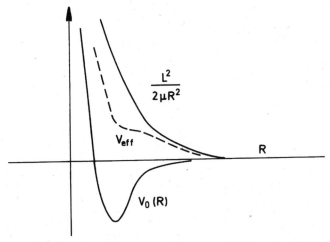

Fig. 3.4. The orbital motion and the interaction potential forms an effective potential.

SUGGESTED READING

H. Goldstein, "Classical Mechanics," Addison Wesley, 1964, Tokyo.

EXERCISES

1. Show the derivation of Eq. (3.12).

2. Using $\mathbf{L} = \mu \mathbf{R} \times \dot{\mathbf{R}}$ show that $L = \mu v b$.

4

COLLISIONAL APPROACH

Traditionally, chemical reactions are characterized by the magnitude and temperature dependence of rate constants. However, a rate constant is a heavily averaged quantity, i.e., averaged over initial quantum states and kinetic energy. During the last 30 years new experimental techniques have made it possible to probe and investigate chemical reactivity at a much more detailed level by being able to measure state to state probabilities or cross sections. In this section these quantities will be defined by the so-called collisional approach to reactive scattering. Using this approach it is possible to calculate (given a suitable potential energy surface) reaction probabilities, cross sections, and rate constants by solving the dynamical equations, using either classical or quantum mechanics. The information on the interaction potential, however, is often not sufficiently detailed to allow such a calculation from "first principles." Therefore it is extremely useful to have simpler approximate methods, i.e., methods that require less information on the interaction potential. Examples of such methods are statistical and transition state methods.

Additionally the solution of the Schrödinger equation for nuclear motion has turned out to be a difficult task, which cannot be carried out routinely. Since the nuclei are quite heavy one might expect that a classical mechanical description of the motion to be valid. This has also turned out to be the case, and for averaged quantities classical dynamics often offers a sufficiently accurate description. Possible exceptions are systems containing light atoms at low temperatures, where quantum mechanical tunneling effects are important. Also for state-to-state resolved cross sections and rate constants the classical dynamical description may fail.

4.1   THE REACTION CROSS SECTION

In order to derive an expression for the cross section we consider a cylinder (see Fig. 4.1) with a cross sectional area σ_{AB} and a length equal to the distance a particle A moves per unit time. Thus if the velocity of particle A relative to B is v, then the flux of particles A that crosses the area σ_{AB} per unit time is $I_A \sigma_{AB}$, where the flux per unit area is

$$I_A(x) = v n_A(x) \qquad (4.1)$$

and $n_A(x)$ is the number density of A particles at a point x. As the particles pass through the cylinder some of them will eventually undergo a collision with B particles. Such collisions will, as we have seen (Fig. 3.2) deflect the particle A out of the beam. Thus the collisions will decrease the flux of A particles passing through dx such that the change is equal to the number of collisions, i.e.:

$$-dI_A = \sigma_{AB}(v)I_A(x)\, n_B \, dx \qquad (4.2)$$

where we have assumed that the number of such collisions will be proportional to the flux of A particles and the number of targets (B particles) in the volume $\sigma_{AB}\, dx$, i.e., proportional to $n_B\sigma_{AB}\, dx$. The cross-sectional area σ_{AB} is a measure of the size of the target.

   The cross section has the dimension of an area and is called the collision cross section. For "hard sphere" molecules the cross section is

$$\sigma = \pi d^2 \qquad (4.3)$$

where d is the sum of the radii of the colliding spheres. These radii are usually taken from the so-called Lennard Jones distances (see Table 4.1). The total number of collisions between A and B molecules per unit time and volume is $n_A n_B v \sigma_{AB}$ and the collision number $Z(v)$ may then be defined as

$$Z(v) = v\sigma_{AB}(v) \qquad (4.4)$$

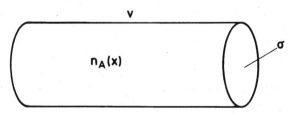

Fig. 4.1. The molecule A with velocity v moves through the cylinder with a cross sectional area σ. $n_A(x)$ is the number density of A particles at point x.

**TABLE 4.1. Lennard Jones Distances for Some
Selected Molecules**

Molecule	d_0 (Å)	Molecule	d_0 (Å)
Ar	3.4	Cl_2	4.2
Kr	3.6	HCl	3.3
H_2	2.9	CH_4	3.8
N_2	3.7	CO_2	3.9–4.2
CO	3.6–3.7	O_2	3.4–3.5

Data are taken from [1]. The Lennard Jones potential is defined
by $V(r) = 4\epsilon[(d_0/r)^{12} - (d_0/r)^6]$, where ϵ is the well depth and
d_0 the value of r, where the potential is zero, i.e., $V(d_0) = 0$.

By averaging over the velocity distribution $f(v, T)$ we obtain the thermal collision number $Z(T)$, i.e.,

$$Z(T) = \int v\sigma_{AB}(v)f(v, T)\, dv \qquad (4.5)$$

where

$$\int f(v, T)\, dv = 1$$

The collision number will, as we can see, depend upon the size of the total
cross section σ_{AB}, which is the effective size of the target B that the molecules
A feel when they approach the target. The cross section will be defined more
specifically below. At present we can introduce a reactive cross section by
demanding that the outcome of the collision should be a reactive event. Thus the
reactive cross section σ_r is defined as the effective size of the target, provided
that we have a reaction, and the corresponding thermally averaged quantity is
a rate constant for the reaction, i.e.,

$$k(T) = \int v\sigma_r(v)f(v, T)\, dv \qquad (4.6)$$

So far we have only accounted for σ's dependence on the relative velocity v.
Molecular-beam experiments, however, also give information on the angular
distribution of the products. This is illustrated (in the impact parameter picture)
in Fig. 4.2. Note that molecules approaching within the impact parameter range
$[b; b + db]$ are scattered with a deflection angle θ with respect to the direction
of the relative motion and with the azimuthal angle ϕ. Thus one would be able
to determine the number of molecules with initial velocity v scattered into a

solid angle $d\omega = \sin\theta \, d\theta \, d\phi$ defined by the range $[\theta; \theta + d\theta]$ and $[\phi; \phi + d\phi]$. Defining the quantity $dN(v, \theta, \phi)$ to be this number we see that the total number of particles scattered (irrespective of the angle) is:

$$N(v) = \int dN(v, \theta, \phi) \, d\omega \qquad (4.7)$$

Likewise we can introduce the differential cross section

$$d\sigma'(v, \theta, \phi) \qquad (4.8)$$

such that the total cross section is obtained as

$$\sigma(v) = \int d\sigma'(v, \theta, \phi) \, d\omega \qquad (4.9)$$

Often the ϕ dependence is either weak or not of interest, e.g., because it may not be probed experimentally. We can then integrate over the ϕ angle to get:

$$\sigma(v) = 2\pi \int d\sigma'(v, \theta) \sin\theta \, d\theta \qquad (4.10)$$

As shown in the Fig. 4.2, there is a relation between the scattering angle θ and the impact parameter range $[b; b + db]$ such that the molecules entering this impact parameter range will be deflected into the range $[\theta; \theta + d\theta]$ and the contribution to the cross section would be $2\pi b \, db$. The total cross section is then

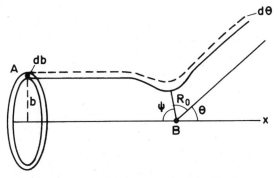

Fig. 4.2. Scattering of a molecule with impact parameter in the range $[b; b + db]$ into the range $[\theta; \theta + d\theta]$.

$$\sigma(v) = \int_{o}^{b_{max}} 2\pi b \; db \qquad (4.11)$$

where b_{max} is the maximum value of b for which scattering occurs. For a hard-sphere collision $b_{max} = d$ (the hard-sphere diameter) and hence $\sigma = \pi d^2$. For reactive scattering one introduces the so-called opacity function $P(v, b)$, which is defined as the fraction of the collisions with a given impact parameter that for a given velocity lead to a reaction. Thus the reactive cross section one obtains is

$$\sigma_r(v) = 2\pi \int_{0}^{b_{max}} P(v, b) b \; db \qquad (4.12)$$

where b_{max} is so large that the probability for reaction vanishes. Instead of integrating over the impact parameter we may integrate over the orbital angular momentum. By using that $L = \mu v b$ we obtain

$$\sigma_r(v) = \frac{\pi}{\mu E} \int_{0}^{L_{max}} P(v, L) L \; dL \qquad (4.13)$$

The scattering angle θ is defined as $\theta = |\xi|$ where $\xi = \pi - \psi$. By integrating Eq. (3.12) using Eq. (3.15) we obtain

$$\xi = \pi - 2b \int_{R_0}^{\infty} \frac{dR}{R^2} \left[1 - \frac{V(R)}{E} - \frac{b^2}{R^2} \right]^{-1/2} \qquad (4.14)$$

where (see Fig. 4.2) R_0 is the turning point for the radial motion and $E = \mu v^2 / 2$, the initial kinetic energy. The integral (4.14) can be evaluated analytically for some simple potentials, and hence the differential cross section $d\sigma'$ can also be evaluated using (from Eqs. (4.10) and (4.11))

$$d\sigma'(v, \theta) = \sum \frac{b}{\sin \theta \left| \dfrac{d\theta}{db} \right|} \qquad (4.15)$$

Since the same scattering angle can be obtained from more than one impact parameter we have introduced a sum over these contributions in Eq. (4.15).

REFERENCES

[1] J. O. Hirschfelder, C. F. Curtiss and R. B. Bird, "Molecular Theory of Gases and Liquids," John Wiley & Sons, New York, 1954.

[2] K. M. Ervin and P. B. Armentrout, J. Chem. Phys. **86**,2659(1987).

SUGGESTED READING

M. S. Child, "Molecular Collision Theory," Academic Press, New York, 1974.

EXERCISES

1. What is the dimensional units for $Z(T)$?

2. Why is Eq. (4.6) a proper definition of a rate constant?

3. Many chemical reactions are dominated by the long-range part of the potential:

$$V(R) = -\frac{C}{R^n} \tag{4.16}$$

where C is a constant and n an integer larger than 2. The effective potential is then:

$$V_{\text{eff}} = -\frac{C}{R^n} + \frac{L^2}{2\mu R^2} \tag{4.17}$$

where L is the orbital angular momentum and μ the reduced mass. Find an expression for the reaction cross section σ_r assuming that the reaction probability $P_r = 1$ for kinetic energies larger than or equal to the barrier in the effective potential.

From the reaction cross section, calculate the rate constant by averaging over a Boltzmann distribution, i.e.,

$$k_r(T) = \sqrt{\frac{8kT}{\pi\mu}} \int_0^\infty d(\beta E_{\text{kin}}) \exp\left(-\beta E_{\text{kin}}\right) \beta E_{\text{kin}} \sigma_r(E_{\text{kin}}) \tag{4.18}$$

where the initial kinetic energy is $E_{\text{kin}} = 0.5\mu v^2$ and $\beta = 1/kT$.

Utilize the definition of the gamma function as:

$$\Gamma(n) = \int_0^\infty \exp(-x)x^{n-1}\,dx \qquad (4.19)$$

For ion-molecule reactions the leading long-range potential can be expressed as

$$V(R) = -\frac{\alpha e^2}{2R^4} \qquad (4.20)$$

where α is the polarizability of the molecule, e the charge of an electron, and $e^2/a_0 = 1$ hartree. Calculate the rate constant for the reaction:

$$N^+ + H_2 \rightarrow NH^+ + H \qquad (4.21)$$

The hydrogen polarizability is $\alpha = 0.79$ Å^3 for hydrogen with a bond length equal to its equilibrium distance ~ 0.74 Å. Experimentally the cross section has been measured to be 15 Å^2 for $E_{kin} = 100$ kJ/mol, 10 Å^2 at 300 kJ/mol and 30 Å^2 at 30 kJ/mol [2]. How well does the simple model above agree with the experimental data?

4. The value of the rate constant, sometimes called 1 GK (gas kinetics), is obtained by assuming that every collision leads to a reaction if the impact parameter is smaller than a hard-sphere diameter.

Calculate the value of 1 GK at 300 K for the collision of H_2 and OH, assuming that $d_0(OH) \sim d_0(H_2) = 2.9$ Å (see Table 4.1).

How does this expression change if we require that the relative kinetic energy should exceed a certain barrier value E_0 before the reaction can occur with unit probability?

Use the expression to calculate the rate constant for the reaction

$$H_2 + OH \rightarrow H_2O + H \qquad (4.22)$$

at 300 K. Use $E_0 = 25.4$ kJ/mol. How does the result compare with the transition state value of $2.9 \cdot 10^{-15}$ cm^3/sec?

5. Derive Eq. (4.14).

6. Calculate the differential cross section, for hard-sphere collisions, where $V(R) = 0$ for $R \geq d$ and $V(R) = 0$ for $R < d$; and for coulomb scattering where $V(R) = a/R$.

5

PARTITION FUNCTIONS

As mentioned previously, reactivity may depend critically upon the initial excitation of the various degrees of freedom of the reacting molecules. The degrees of freedom, listed in order of increasing level spacing, are translational, rotational, vibrational, and electronic. When molecules collide, energy is transferred from one degree of freedom to the other by either intramolecular or intermolecular energy transfer. If a molecule is initially excited in some degree of freedom the energy will be redistributed by inter- and intramolecular energy transfer in a number of collisions. The number of collisions needed for this equilibration depends upon the degree of freedom in question, since translational energy is redistributed faster than rotational energy and so on. (Exceptions may occur if the interaction allows for resonance transitions between the two molecules). Often we will then be able to assume an equilibrium distribution with respect to the translational energy and also the rotational energy, since these degrees of freedom "equilibrate" rapidly (after a few collisions). The vibrational energy on the other hand will usually only be equilibrated on a much longer time scale, and for the electronic degrees of freedom we often only have to consider the ground state. This separation of the various degrees of freedom on the energy scale is fortunate, since it allows us to assume an equilibrium distribution of some degrees of freedom while studying nonequilibrium effects (such as chemical reactions) in others. It must always be kept in mind, however, that we are then making an assumption, and many discrepancies between theoretically and experimentally determined rate constants have been due simply to an unjustified assumption of equilibrium. For a brief summary of relevant statistical mechanics see Appendix D.

The equilibrium system is characterized by a particular temperature T, and the energy distribution follows Boltzmann's distribution law. Thus the fraction of molecules in a given quantum state i is given as:

$$\frac{N_i}{N} = \frac{\exp(-\epsilon_i/kT)}{\sum_i \exp(-\epsilon_i/kT)} \tag{5.1}$$

where ϵ_i is the energy of the state and k is Boltzmann's constant. If the state i is degenerate, the expression still holds if i runs over degenerate states as well. However we usually introduce the degeneracy factor g_i and the sum then runs over states with different energies, i.e.,

$$\frac{N_i}{N} = \frac{g_i \exp(-\epsilon_i/kT)}{Q} \tag{5.2}$$

where Q is the partition function

$$Q = \sum_i g_i \exp(-\epsilon_i/kT) \tag{5.3}$$

Thus in order to evaluate the partition function for a given degree of freedom we need to know the degeneracy factor and the energy level. Both quantities are available from quantum theoretical solutions of the Schrödinger equation for the molecule. Here we shall restrict ourselves to a consideration of simple model systems for which the problem is solvable analytically. The models are:

For **translation:** The particle in a box (no interaction potential)
For **rotation:** A rigid rotor
For **vibration:** The one dimensional (1-D) harmonic oscillator

The energy levels and degeneracies for these model systems are given in Table 5.1. In Table 5.1, h is Planck's constant, j the rotational angular momentum, and I the moment of inertia. For the particle in a box (model for the translational motion) we then obtain the partition function:

$$Q_t = \sum \frac{dx\,dp_x}{h} \exp\left(-\frac{p_x^2}{2mkT}\right) \tag{5.4}$$

The subscript t indicates the translational model. Replacing the summation

TABLE 5.1. Energy Levels and Degeneracy Factors

Model	g_i	ϵ_i
Particle in box (1-D)	$\dfrac{dx\,dp_x}{h}$	$\dfrac{p_x^2}{2m}$
Particle in box (3-D)	$\dfrac{dx\,dy\,dz\,dp_x\,dp_y\,dp_z}{h^3}$	$\dfrac{p_x^2 + p_y^2 + p_z^2}{2m}$
Rotation (diatomic)	$2j + 1$	$\dfrac{\hbar^2 j(j+1)}{2I}$
Vibration (1-D harmonic oscillator)	1	$(v + \tfrac{1}{2})\,\hbar\omega$

(over all states) by integrals and using a box of length L_x we find

$$Q_t = \frac{\sqrt{2\pi mkT}}{h}\, L_x \qquad (5.5)$$

Thus the fraction of molecules with momentum in the interval $[p_x, p_x + dp_x]$ irrespective of x is

$$\frac{dN(p_x)}{N} = \frac{1}{Q_t}\,\exp(-p_x^2/2mkT)\,\frac{dp_x}{h}\int_0^{L_x} dx \qquad (5.6)$$

which becomes

$$\frac{dN(p_x)}{N} = \sqrt{\frac{1}{2\pi mkT}}\,\exp(-p_x^2/2mkT)\,dp_x \qquad (5.7)$$

Considering the 3-D case we get, in a similar fashion,

$$\frac{dN(p_x p_y p_z)}{N} = \left(\frac{1}{2\pi mkT}\right)^{3/2}\exp(-p^2/2mkT)\,dp_x\,dp_y\,dp_z \qquad (5.8)$$

where $p^2 = p_x^2 + p_y^2 + p_z^2$. Often we are interested in the distribution function for p, the magnitude of the momentum, irrespective of the direction of **p**. This is found by introducing the polar angles θ, ϕ such that

$$p_x = p \sin(\theta) \cos(\phi) \qquad (5.9)$$

$$p_y = p \sin(\theta) \sin(\phi) \qquad (5.10)$$

$$p_z = p \cos(\theta) \qquad (5.11)$$

i.e.,

$$dp_x \, dp_y \, dp_z = p^2 \, dp \sin(\theta) \, d\theta \, d\phi \qquad (5.12)$$

and integrating over the polar angles. Thus we obtain

$$\frac{dN(p)}{N} = 4\pi p^2 \left(\frac{1}{2\pi m k T} \right)^{3/2} \exp(-p^2/2mkT) \, dp \qquad (5.13)$$

Using this distribution function we obtain the average momentum

$$\langle p \rangle = \frac{\int_0^\infty p \, dN(p)}{N} = \sqrt{\frac{8kTm}{\pi}} \qquad (5.14)$$

In the same manner we obtain the average kinetic energy

$$\left\langle \frac{p^2}{2m} \right\rangle = \frac{3}{2} kT \qquad (5.15)$$

Thus we have

$$\left\langle \frac{p_x^2}{2m} \right\rangle = \left\langle \frac{p_y^2}{2m} \right\rangle = \left\langle \frac{p_z^2}{2m} \right\rangle = \frac{1}{2} kT \qquad (5.16)$$

which is a manifestation of the equipartition theorem, i.e., the energy in each independent degree of freedom is $kT/2$.

We have listed the partition functions for other model cases in Table 5.2.

TABLE 5.2. Molecular Partition Functions

Type of Energy	Partition Function
Translational energy in 1-D	$Q_t = \dfrac{\sqrt{2\pi mkTL}}{h}$
Translational energy in 3-D	$Q_t = \dfrac{(2\pi mkT)^{3/2}V}{h^3}$
Rotational energy of linear molecule with moment of inertia I	$Q_r = \dfrac{2IkT}{\sigma \hbar^2}$
Rotational energy of nonlinear molecule with principal moments of inertia I_a, I_b, I_c	$Q_r = \dfrac{\sqrt{8\pi I_a I_b I_c}(kT)^{3/2}}{\sigma \hbar^3}$
Vibrational energy of harmonic oscillator relative to zero point energy level	$Q_v = \dfrac{1}{1 - \exp(-\hbar\omega/kT)}$
Vibrational energy of s vibrational degrees of freedom relative to zero point energy level	$Q_v = \Pi_i (1 - \exp(-\hbar\omega_i/kT))^{-1}$
Electronic energy level (widely spaced)	$Q_e = g_e$ (degeneracy factor of ground state)

σ is the symmetry number = no. of equivalent arrangements.

5.1 EQUILIBRIUM CONSTANTS

According to statistical mechanics (Appendix D) the chemical potential μ_i for species "i" can be expressed in terms of the partition function as:

$$\mu_i = -kT \ln(Q_i/N_i) \qquad (5.17)$$

where N_i is the number of molecules i. Introducing the equilibrium condition

$$\sum_i \nu_i \mu_i = 0 \qquad (5.18)$$

where ν_i is the stoichiometric coefficient (negative for reactants and positive for products), the equation may be written as:

$$\Pi_i N_i^{\nu_i} = \Pi_i Q_i^{\nu_i} \exp(-\Delta E_0/kT) \qquad (5.19)$$

Introducing the volume V and dividing by $V^{\Sigma \nu_i}$ we get:

$$\Pi_i \left(\frac{N_i}{V} \right)^{\nu_i} = \Pi_i \left(\frac{Q_i}{V} \right)^{\nu_i} \exp(-\Delta E_0/kT) \qquad (5.20)$$

where the left-hand side is the equilibrium constant and ΔE_0 is the energy difference between the zero point energies of the products and reactants. Thus the partition functions are calculated using the ground state vibrational level as the zero of the energy scale (see Table 5.2), both for reactants and products.

The partition function is

$$Q_i = Q_t Q_{\text{int}} \qquad (5.21)$$

a product of a translational partition function, Q_t, and one, Q_{int}, representing the internal degrees of freedom of molecules i.

SUGGESTED READING

F. Reif, "Fundamentals of Statistical and Thermal Physics," McGraw-Hill, New York, 1965.

EXERCISES

1. Calculate the average speed using a one-dimensional distribution function.

2. Find the rotational partition function for a diatomic molecule.

3. Introduce expression (5.13) in Eq. (4.6) to get

$$k(T) = \sqrt{8kT/\pi m} \int_0^\infty d(\beta E_{\text{kin}}) \beta E_{\text{kin}} \exp(-\beta E_{\text{kin}}) \sigma_r(E_{\text{kin}}) \qquad (5.22)$$

4. Given that the energy levels for the particle in a box are

$$\epsilon_{n_x} = \frac{\hbar^2}{2m} k_x^2 \qquad (5.23)$$

where $k_x = \pi n_x/L$, $n_x = 1, 2, \ldots$, derive Eq. (5.5) and show that the degeneracy factor is $g = dx\, dp_x/h$.

6

TRANSITION STATE THEORY

Transition state theory allows one to obtain the reaction rate constant from the molecular properties of the reactants and the ones corresponding to the "transition state." The latter state is defined as an activated complex from which the system will, once it is reached, form products. Thus we only have to calculate the concentration of the activated complex $[AB^{\#}]$ and the rate with which the system goes through the transition state configuration (denoted by #). Thus the process is considered to be

$$A + B \rightleftharpoons AB^{\#} \rightarrow \text{products} \tag{6.1}$$

where A and B are atoms or molecules. The transition state is conventionally taken to be positioned at the top of the activation barrier (see Fig. 6.1). Mathematically it is the point on the potential energy surface where the energy gradient, with respect to the nuclear coordinates, is zero and the energy hessian has one negative eigenvalue (the energy hessian is the second derivative matrix with respect to the nuclear coordinates). The formation of the activated complex is assumed to take place through an equilibrium with the reactants and hence an equilibrium constant can be introduced

$$\frac{[AB^{\#}]}{[A][B]} = K^{\#} \tag{6.2}$$

A more precise way of describing the motion from reactants to the activated complex and to products is to introduce the so-called reaction path, in which one degree of freedom is projected out, which then for a molecule $AB^{\#}$ with

31

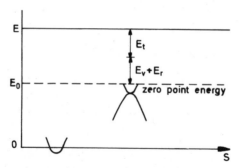

Fig. 6.1. Energetics at the transition state. E_0 is the effective barrier including zero-point vibrational effects.

N atoms leaves 3N–7 modes for the vibrational degrees of freedom (see Table 6.1). The motion along the reaction path is then a one-dimensional translation that brings the activated complex to products. At this point we introduce the essential assumption of transition state theory, namely that of no recrossing. Thus once the system has crossed the transition state (a point along the reaction path) it does not return. With this assumption we see that the reaction rate r is

$$r = [AB^{\#}]\nu \tag{6.3}$$

where ν is a frequency associated with the passage over the transition state. Substituting Eq. (6.2) we get

$$r = \nu K^{\#}[A][B] \tag{6.4}$$

and hence the reaction rate constant, k_R, obtained is:

$$k_R = \nu K^{\#} \tag{6.5}$$

TABLE 6.1. Degrees of Freedom for the Activated Complex and Reactants[a]

Degrees of Freedom	A	B	$AB^{\#}$
Translation	3	3	3
Rotation	3	3	3
Vibration	$3N_A-6$	$3N_B-6$	$3N-7$
Reaction path motion			1
Total	$3N_A$	$3N_B$	$3N$

[a]Nonlinear molecules.

The passage frequency can be written as

$$\nu = \frac{\langle v^{\#} \rangle}{\Delta s} \tag{6.6}$$

where $\langle v^{\#} \rangle$ is the average speed and Δs a distance along the reaction path s. The average speed in the forward direction is given by (Exercise 4, Chapter 5):

$$\langle v^{\#} \rangle = \sqrt{\frac{kT}{2\pi m_{AB}}} \tag{6.7}$$

where m_{AB} is a mass connected to the motion of the complex along the reaction path (its precise definition is not important here since it vanishes later on). The equilibrium constant $K^{\#}$ can be evaluated from the partition functions as:

$$K^{\#} = \frac{g_e^{\#} Q_v^{\#} Q_r^{\#} Q_t^{\#} Q_t^{*}}{g_e^A Q_v^A Q_r^A Q_t^A g_e^B Q_v^B Q_r^B Q_t^B} \, V \exp(-\Delta E_0/kT) \tag{6.8}$$

where we have used the fact that the spacing between the electronic states is usually sufficiently large that we can replace the partition functions with the electronic degeneracy factor for the reactant and transition state complex, and that the partition function is given by a product of the partition functions for the different types of motion.

The partition function for the one-dimensional translational motion of the complex along the reaction path is given by (the * indicates that it is along the reaction path):

$$Q_t^{*} = \frac{\sqrt{2\pi m_{AB} kT}}{h} \Delta s \tag{6.9}$$

Before substituting the above expressions into Eq. (6.5) we must consider the energy scale on which we work. The total energy of the system E can be partitioned among the various degrees of freedom:

$$E = E_t + E_r + E_v + E_e \tag{6.10}$$

Since we usually calculate the vibrational partition function (see Table 5.2) excluding the zero point vibrational energy, the expression for ΔE_0 is

$$\Delta E_0 = \frac{1}{2} \sum_{i=1}^{3N-7(3N-6)} \hbar\omega_i^{\#} + V(s^{\#})$$
$$- \frac{1}{2} \sum_{i=1}^{3N_A-6(3N_A-5)} \hbar\omega_i^A - \frac{1}{2} \sum_{i=1}^{3N_B-6(3N_B-5)} \hbar\omega_i^B - V(s = -\infty) \quad (6.11)$$

where $V(s^{\#})$ and $V(s = -\infty)$ indicate the values of the potential (electronic) energy at the transition state and of the reactants respectively (see Fig. 6.1). The numbers in parentheses in the summations hold for linear molecules and linear transition state configuration.

Finally we obtain the following expression for the reaction rate constant

$$k_R = \sigma_{\text{symm}} \frac{kT}{h} \frac{Q^{\#}}{Q_A Q_B} \exp(-\Delta E_0/kT) \quad (6.12)$$

where the electronic degeneracy factor g_e has been set to unity and σ_{symm} is a common symmetry factor, i.e., the number of identical reaction paths. The partition function can be expressed as a product of partition functions for the translational motion and the internal motion (rotation and vibration), i.e.,

$$Q^{\#} = Q_t^{\#} V Q_r^{\#} Q_v^{\#}, \quad (6.13)$$
$$Q_A = Q_t^A Q_r^A Q_v^A, \quad (6.14)$$
$$Q_B = Q_t^B Q_r^B Q_v^B, \quad (6.15)$$

where V is the volume. Thus the rotational partition functions are **without** symmetry factors. They have all been combined in the more rigorous counting of the number of equivalent paths through σ_{symm}. Notice also that the ratio $Q_t^A Q_t^B / V Q_t^{\#}$ may be written as:

$$\tilde{Q}_t = \frac{(2\pi\mu kT)^{3/2}}{h^3} \quad (6.16)$$

where μ is the reduced mass and $\tilde{Q}_t$ is the translational partition function per unit volume.

As mentioned, transition state theory does not account for the possibility of recrossing the transition state. In order to compensate for this assumption in the theory, one multiplies the transition state rate constant with a so-called transmission factor κ. From these considerations one would expect that transition state theory would necessarily overestimate the rate constant. However, the transmission factor may take on values larger than unity. This can occur for example when tunneling is important. Tunneling through an activation barrier is a quantum mechanical effect not accounted for in transition state theory. Thus the transmission factor may be larger than, equal to, or smaller than unity. How to define this correction factor is considered in the next chapter. First we shall apply transition state theory to a simple example, namely that of a collinear collision of an H atom and a chlorine molecule Cl_2, i.e., to consider the reaction

$$H + Cl_2 \rightarrow HCl + Cl \tag{6.17}$$

Note that the evaluation of a rate constant using transition state theory requires information about the frequencies for the vibrational motion and the moments of inertia of the complex. In some cases, for example the one above, one may argue as follows: From the model potential energy surface (LEPS) we know that the barrier is 0.1076 eV and in a collinear model the rotational partition functions are omitted. Furthermore the bending vibration of the $H \cdots Cl—Cl$ complex can be neglected. Let us furthermore assume that the vibrational motion of the complex resembles that of Cl_2 (due to the fact that we have an early barrier and that the H-atom is light compared to Cl). Thus $Q_v^{\#} \sim Q_v^{Cl_2}$ and we are left with just the translational partition function, which for one-dimensional motion is

$$Q_t = \frac{\sqrt{2\pi m_H kT}}{h} \tag{6.18}$$

where we have replaced the reduced mass by m_H. Thus the transition state rate constant becomes (σ_{symm} is set to unity)

$$k(T) = \sqrt{\frac{kT}{2\pi m_H}} \, \exp\left(-\Delta E_0 / kT\right) \tag{6.19}$$

In Table 6.2 the values obtained by using this simple model are compared with exact quantum calculations. One notes that the transmission factor is larger than unity at low temperatures; thus some tunneling is not accounted for by transition state theory. The isotopic ratio would, in the transition state model, be predicted to be

$$k(H + Cl_2)/k(D + Cl_2) = \sqrt{m_D/m_H} = 1.4 \tag{6.20}$$

At lower temperatures the exact rate constant for the reaction ($H + Cl_2$) is some-

TABLE 6.2. Comparison of Transition State and Quantum Mechanical Rate Constants for the Reaction H + $Cl_2 \rightarrow$ HCl + Cl

Temperature K	Quantum k(T) in	Transition State cm/sec	κ	Isotopic Ratio
200	210	99.6	2.1	1.58
300	1470	977	1.5	1.45
400	4180	3200	1.3	1.41
500	8050	6690	1.2	1.40
600	12600	11100	1.1	1.40

The isotopic ratio between the quantum mechanically obtained rates for hydrogen and deuterium is also given. κ is the transmission factor, i.e., the correction to transition state theory. Data from [1]

what larger than that predicted by transition state theory. This is also in agreement with a contribution from tunneling.

REFERENCES

[1] H. Essen, G. D. Billing and M. Baer, Chem. Phys. **17**,443(1976).

EXERCISES

1. Using transition state theory, calculate the rate constant for the reaction

$$H_2 + OH \rightarrow H_2O + H \tag{6.21}$$

at 200 K, 300 K, 500 K, and 1000 K. The vibrational frequencies for the reactants are $\omega_{H_2} = 4260$ cm^{-1} and $\omega_{OH} = 3620$ cm^{-1}. The collision complex (the transition state) has vibrational frequencies $\omega_1 = 575$ cm^{-1}, $\omega_2 = 832$ cm^{-1}, $\omega_3 = 861$ cm^{-1}, $\omega_4 = 1927$ cm^{-1}, and $\omega_5 = 3558$ cm^{-1}; and moments of inertia $I_1 = 0.6349$ amu Å^2, $I_2 = 5.670$ amu Å^2, and $I_3 = 6.514$ amu Å^2. The activation barrier is $E_0 = 25.4$ kJ/mol, and the equilibrium bond lengths of H_2 and OH are 0.756 Å and 0.986 Å, respectively. Find also the average reaction cross section

$$\langle \sigma_r \rangle = k_r(T) / \sqrt{8kT/\pi m} \tag{6.22}$$

and the average probability for reaction

$$\langle P_r \rangle = \langle \sigma_r \rangle / \sigma_0 \tag{6.23}$$

where σ_0 is the hard-sphere cross section.

2. Consider the chemical reaction

$$CH_4 + H \rightarrow CH_3 + H_2 \qquad (6.24)$$

The normal mode frequencies for the reactants and the transition state complex are given in Table 6.3. The potential barrier is 14.0 kcal/mol, and the electronic degeneracy factor $g_e = 1$.

TABLE 6.3. Vibrational Frequencies for CH_4 and $CH_3 \cdots H \cdots H$ in cm^{-1}

	Degeneracy		
	Single	Double	Triple
CH_4			3167
	3047		
		1573	
			1367
$CH_3 \cdots H \cdots H$		3236	
	3090		
	1760		
		1459	
		1152	
	1111		
		534	

a. Calculate the effective barrier, for which the zero point energy difference is included.

b. What is the value of the symmetry factor σ_{symm}?

c. Give an expression for the reaction rate constant using transition state theory.

d. Calculate the reaction rate constant at 300 K and 1000 K. The rotational partition functions can be neglected.

e. Experimentally one finds that at 300 K, $k = 8.1 \cdot 10^{-20}$ cm^3/sec and at 1000 K, $k = 1.8 \cdot 10^{-13}$ cm^3/sec. Give possible explanations for any differences between theoretical and experimental values.

f. Consider the reaction

$$D + CH_4 \rightarrow CH_3 + HD$$

and discuss where isotope effects enter the calculation.

TABLE 6.4. Vibrational Frequencies for CH$_3$Cl and Cl $\cdots$ CH$_3$ $\cdots$ Cl in cm^{-1}

	Degeneracy	
	Single	Double
CH$_3$Cl		3042
	2933	
	1463	
		1403
		1013
	736	
Cl $\cdots$ CH$_3$ $\cdots$ Cl	3430	
		3274
		1379
		1015
	1004	
	195	188
	i442	

where "i" is the imaginary unit.

3. The gas phase S$_N$2 reaction

$$Cl^- + CH_3Cl \rightarrow ClCH_3 + Cl^- \tag{6.25}$$

has a barrier of 14.9 kJ/mol. The principal moments of inertia are $I_a = I_b$ =38.1 amu Å^2, and $I_c = 3.08$ amu Å^2 for CH$_3$Cl; and $I_a = I_b = 405$ amu Å^2, and $I_c = 3.46$ amu Å^2 for the transition state complex. What is the geometry of the transition state, and what is the symmetry factor σ_{symm}? Find the rate constant at 300 K and 1000 K using transition state theory.

7

GENERALIZED TRANSITION STATE THEORY

Considering the rate constant for a collinear A + BC reaction we have

$$A + BC(n) \rightarrow AB(n') + C \tag{7.1}$$

where n denotes the vibrational state of the diatomic molecule and n' the product vibrational state. The expression for the rate constant is, according to the collisional approach, given as

$$k_n(T) = \int v f(v; T) P_n^r(v) \, dv \tag{7.2}$$

where v is the relative velocity and $f(v; T)$ the one-dimensional distribution of velocities, i.e., (see Chapter 5)

$$f(v; T) = \mu \sqrt{\frac{1}{2\mu\pi k T}} \exp\left(-\frac{1}{2} \mu v^2 / k T\right) \tag{7.3}$$

Here μ is the reduced mass

$$\mu = \frac{m_A m_{BC}}{m_A + m_{BC}} \tag{7.4}$$

$P_n^r(v)$ is the reaction probability for BC molecules in a specific internal vibrational quantum state (n). The expression (7.2) is exact; the only assumption is that the kinetic energy of the relative motion is distributed according to a Maxwell-Boltzmann distribution law. We now introduce the partition function (per unit length) for the translational motion, i.e.,

$$Q_{tr}^{rel}(T) = \frac{\sqrt{2\pi\mu kT}}{h} \tag{7.5}$$

where h is Planck's constant and the superscript "rel" indicates that the partition function is for the relative translational motion. Note that this partition function may be written as:

$$Q_{tr}^{rel}(T)^{-1} = \frac{Q_{tr}^{\#}}{Q_{tr}^{A} Q_{tr}^{BC}} \tag{7.6}$$

where the translational partition functions of the transition state (#) and the reactants have been introduced. From Eqs. (7.2) and (7.5) we then get

$$k_n(T) = \frac{kT}{h} \frac{Q_{tr}^{\#}}{Q_{tr}^{A} Q_{tr}^{BC}} \overline{P}_n^r(T) \tag{7.7}$$

where

$$\overline{P}_n^r(T) = \int d(E_{kin}\beta) \exp\left(-E_{kin}\beta\right) P_n^r(v) \tag{7.8}$$

is the average reaction probability of BC molecules in vibrational state n. E_{kin} is the kinetic energy

$$E_{kin} = \tfrac{1}{2}\mu v^2 \tag{7.9}$$

and $\beta = 1/kT$. We define the total reaction rate constant as a Boltzmann weighted average over the rates for the individual vibrational levels, i.e.,

$$k(T) = \frac{\displaystyle\sum_n \exp\left(-\beta E_n\right) k_n(T)}{\displaystyle\sum_n \exp\left(-\beta E_n\right)}$$

$$= \frac{kT}{h} \frac{Q_{tr}^{\#}}{Q_{tr}^{A} Q_{tr}^{BC} Q_{vib}^{BC}} \sum_n \exp\left(-\beta E_n'\right) \overline{P}_n^r(T) \tag{7.10}$$

where E_n' denote the vibrational energies of reactant diatomic molecule BC relative to the zero point vibrational energy (see Eq. (7.13)), $E_n = E_n(-\infty)$ and

$$Q_{\mathrm{vib}}^{BC} = \sum_n \exp\left[-\beta E_n'(-\infty)\right] \tag{7.11}$$

We now introduce the "activation" barrier for the vibrational state n (see Fig. 7.1)

$$E_n^b = E_n(s_n^*) - E_n(-\infty) \tag{7.12}$$

By introducing the relative vibrational energy (relative to level 0)

$$E_n'(-\infty) = E_n(-\infty) - E_0(-\infty) \tag{7.13}$$

we see that

$$\begin{aligned}
E_n'(-\infty) + E_n^b &= E_n(s_n^*) - E_0(-\infty) \\
&= E_0^b + E_n(s_n^*) - E_0(s_0^*) \\
&= E_0^b + E_n'^b
\end{aligned} \tag{7.14}$$

where we have used Eq. (7.12) with ($n = 0$) and defined $E_n'^b$ as

$$E_n'^b = E_n(s_n^*) - E_0(s_0^*) \tag{7.15}$$

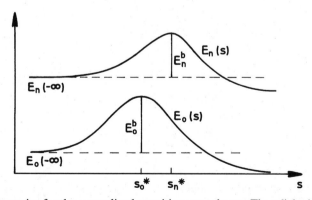

Fig. 7.1. Energetics for the generalized transition state theory. The adiabatic vibrational energy curves along the reaction path are $E_n(s)$.

Thus the Boltzmann factor $\exp(-\beta E'_n(-\infty))$ can, according to Eq. (7.14), be expressed as

$$\exp(-\beta E'_n(-\infty)) = \frac{\exp(-\beta E_0^b)\exp(-\beta E'^b_n)}{\exp(-\beta E_n^b)} \tag{7.16}$$

The expression E'^b_n is the difference between the activation barriers for state n and state 0. Inserting this in Eq. (7.10) we get

$$k(T) = \frac{kT}{h}\frac{Q_{tr}^{\#}}{Q_{tr}^A Q_{tr}^{BC} Q_{vib}^{BC}} \exp(-\beta E_0^b) \sum_n \frac{\exp(-\beta E'^b_n)\overline{P}^r_n(T)}{\exp(-\beta E_n^b)} \tag{7.17}$$

If the dynamics were treated using classical mechanics then the reaction probability would, in this one-dimensional model, be zero below the barrier and unity above it. In other words we can introduce a "classical" (indicated by the superscript cl) reaction probability:

$$P_n^{cl}(E_{kin}) = 0 \quad \text{for} \quad E_{kin} \le E_n^b \quad \text{and} \quad P_n^{cl}(E_{kin}) = 1 \quad \text{for} \quad E_{kin} > E_n^b \tag{7.18}$$

The Boltzmann averaged classical probability is then

$$\overline{P}_n^{cl}(T) = \int d(\beta E_{kin})\exp(-\beta E_{kin})P_n^{cl}(E_{kin}) = \exp(-\beta E_n^b). \tag{7.19}$$

Thus we can express the total reaction rate constant as

$$k(T) = \kappa^{ex}(T)\frac{kT}{h}\frac{Q_{tr}^{\#}Q_{vib}^{\#}}{Q_{tr}^A Q_{tr}^{BC} Q_{vib}^{BC}}\exp(-\beta E_0^b) \tag{7.20}$$

where

$$Q_{vib}^{\#} = \sum_n \exp(-\beta E'^b_n) \tag{7.21}$$

is a partition function (a Boltzmann sum) involving the internal energies measured at the position of the barrier along s (the reaction coordinate). The first factor in Eq. (7.20), usually called the transmission factor, has now an exact

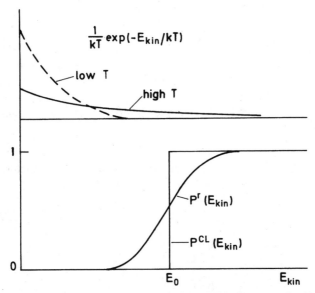

$$\frac{1}{kT}\exp(-E_{kin}/kT)$$

low T

high T

$P^r(E_{kin})$

$P^{CL}(E_{kin})$

Fig. 7.2. The upper panel shows the one-dimensional Boltzmann-distribution of the kinetic energy at high and low temperature. The lower panel shows the classical (P^{CL}) and quantum (P^r) reaction probabilities as a function of kinetic energy.

interpretation. It is given by

$$\kappa^{ex}(T) = \frac{1}{Q_{vib}^{\#}} \sum_n \exp(-\beta E_n'^b) \frac{\overline{P}_n^r(T)}{\overline{P}_n^{cl}(T)} \qquad (7.22)$$

We emphasize that the final expression for the rate constant is exact, i.e., no assumptions about equilibrium between the collision complex ABC and the reactants have been made. Note also that the transmission factor has been defined as a Boltzmann average (averaged over saddle point energies) of the ratio of the exact (i.e., quantum mechanical) and classical mechanical average reaction probabilities. The transmission factor may be larger than, equal to, or smaller than one [1, 2]. At low temperatures the average quantum reaction probability will be larger than the classical due to tunneling. At high temperatures the opposite is the case, due to reflection above the barrier (see Fig. 7.2).

REFERENCES

[1] A. Kuppermann, J. Phys. Chem. **83,**171(1979).

[2] H. Eyring, J. Walter and G. E. Kimball, "Quantum Chemistry," John Wiley & Sons, New York, 1965.

EXERCISE

1. Calculate the transmission factor for the reaction $H + Cl_2 \rightarrow HCl + Cl$ at 200 K, 300 K, and 500 K. Since the reaction has an early barrier we can assume that $E_n^{'b} = E_n'(-\infty)$ and $E_n^b = E_0^b = 0.107$ eV. For Cl_2 we have $E_n(-\infty)$ = 0.035 eV, 0.103 eV, and 0.171 eV for ($n = 0, 1, 2$, respectively). Use the data in Table 7.1 and, for example, Simpson's rule to integrate over the kinetic energy.

TABLE 7.1. Quantum Reaction Probabilities as a Function of Kinetic Energy and Initial Quantum State

E_{kin}(eV)	$P_0^r(v)$	$P_1^r(v)$	$P_2^r(v)$
0.065	0.012	0.020	0.035
0.075	0.045	0.11	0.09
0.085	0.133	0.24	0.20
0.095	0.32	0.46	0.39
0.105	0.57	0.73	0.62
0.115	0.79	0.86	0.81
0.125	0.90	0.93	0.91
0.135	0.96	0.97	0.95
0.145	0.98	0.98	0.96
0.155	0.98	0.98	0.95
0.165	0.99	0.99	0.95

$P_n^r(v)$ is the reaction probability from quantum state n of the Cl_2 molecule at the relative velocity v.

8

THEORY FOR UNIMOLECULAR REACTIONS

Unimolecular reactions occur as suggested by Lindemann [1] by succesive excitation of a molecule A through collisions with bath molecules M to an energized state A^* from which the molecule has a probability (different from zero) for undergoing a unimolecular reaction, i.e., to form products

$$A + M \underset{k_2}{\overset{k_1}{\rightleftharpoons}} A^* + M \tag{8.1}$$

$$A^* \overset{k_a}{\rightarrow} products \tag{8.2}$$

The rate equation for the disappearance of A molecules can be obtained using a steady-state treatment, where the concentration of A^* is assumed to be stationary. Thus using that $d[A^*]/dt \sim 0$ we get:

$$-\frac{d[A]}{dt} = \frac{k_1 k_a [A][M]}{k_2[M] + k_a} = k_{uni}[A] \tag{8.3}$$

where we have introduced a "unimolecular" rate constant k_{uni}:

$$k_{uni} = \frac{k_1 k_a}{k_2 + k_a/[M]} \tag{8.4}$$

We can see that the expression, at high pressures, approaches a constant $k^\infty =$

45

$k_1 k_a / k_2$, and at low pressures is pressure dependent, i.e.,

$$k_{uni} \sim k^0 = k_1 [M] \tag{8.5}$$

(∞ and 0 symbolize high and low pressure, respectively). Thus we can introduce the reduced quantity k_{uni}/k^{∞} as:

$$\frac{k_{uni}}{k^{\infty}} = \frac{k^0}{k^0 + k^{\infty}} \tag{8.6}$$

This ratio interpolates between the high- and low-pressure limits. Although the expression in Eq. (8.6) is qualitatively correct it does not represent the experimentally determined "fall off" region accurately (see Fig. 8.1). In order to understand why this difference between the simple Lindemann model and experiment exists it is necessary to consider the activation process in more detail. Early theories of Hinshelwood, Rice, Ramsperger, Kassel [2] and Slater [3] assumed activation of the molecule to occur through the vibrational degrees of freedom treating these degrees of freedom as classical or quantum oscillators. Later the foundation of a microscopic rate theory (known as the RRKM theory) was laid, primarily by Marcus [4]. In the RRKM theory one distinguishes between an energized state, from which reaction usually does not take place. This state is denoted by (*). The activated complex, from which the unimolecular reaction can take place is denoted by (#). Here we have for simplicity omitted this distinction, i.e., introduced the assumption that the reaction rate between them is much larger than the unimolecular decay rate.

In all the above-mentioned theories the rates k_a, k_1, and k_2 are assumed to be dependent on the energy E of the molecule, and it is assumed that dissociation

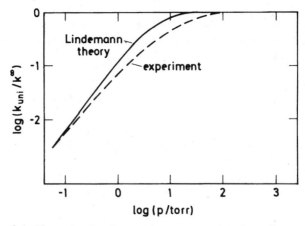

Fig. 8.1. The unimolecular reaction rate as a function of pressure.

can occur if this energy exceeds a critical value E_0. Thus from Eq. (8.4) we can write

$$k_{uni} = \sum_{E \geq E_0} \frac{k_a(E) \dfrac{k_1(E)}{k_2(E)}}{1 + \dfrac{k_a(E)}{k_2(E)[M]}} \tag{8.7}$$

where we have indicated that the rates depend on the energy E and where $k_1(E)/k_2(E)$ is the fraction of molecules having energy in the range $[E; E+dE]$. If this ratio is introduced as $P(E,T)\,dE$ we have

$$k_{uni}(T) = \int_{E_0}^{\infty} dE\, \frac{k_a(E)P(E,T)}{1 + \dfrac{k_a(E)}{k_2(E)[M]}} \tag{8.8}$$

More elegant expressions introduce additionally the total angular momentum dependence of the "microscopic" rates. However, the correct dynamical solution of the problem of activation seeks the solution through a so-called Master equation. The excitation of a molecule to an activated state, from which dissociation can take place, occurs through collisions with bath molecules. This gas could eventually consist of the A molecules themselves. The collisions excite the internal degrees of freedom (rotation and vibration) of the molecule. Thus if we consider the level population $n_i(t)$ of level i at time t we can introduce a set of equations describing the level population as

$$\frac{dn_i}{dt} = [M] \sum_{j} (n_j k_{ji} - n_i k_{ij}) - k_i n_i \tag{8.9}$$

where i denotes a vibrational/rotational level and where k_{ij} are the rates for transitions from level i to level j, and k_i the rate constant for reactions from level i (see Fig. 8.2). Here we assume that only levels above a certain critical energy E_0 can react, i.e., have k_i different from zero. Associated with the above equation (the Master equation) are two problems that prevent its solution in most cases. The first problem is that the rate constants for the transitions among the levels of the molecule are in general not known. Secondly the number of levels, i, is enormous: of the order of 10^{15} states per cm^{-1} is not atypical. This fact justifies replacement of the summation by an integration over energy, i.e.,

$$\frac{\partial n(E,t)}{\partial t} = [M] \int_{0}^{\infty} [k(E',E)n(E',t) - k(E,E')n(E,t)]\,dE' - k(E)n(E,t) \tag{8.10}$$

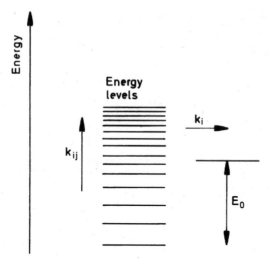

Fig. 8.2. Collisional excitation up the energy ladder of a molecule. Unimolecular decay can occur for energies above the barrier E_0.

If we now assume (for the sake of simplicity) that the level population $n(E,t)$ undergoes an exponential decay, such that

$$n(E,t) = n(E) \exp(-k_{uni}t) \qquad (8.11)$$

we obtain

$$-k_{uni}n(E) = [M] \int_0^\infty [k(E',E)n(E') - k(E,E')n(E)]\, dE' - k(E)n(E) \qquad (8.12)$$

or by integrating over the energy E:

$$k_{uni} = \frac{\int dE\, k(E)n(E)}{\int dE\, n(E)} - [M]\frac{\int dE \int dE'\, [k(E',E)n(E') - k(E,E')n(E)]}{\int dE\, n(E)} \qquad (8.13)$$

We shall now consider two limiting cases, namely the high- and low-pressure limits.

8.1   THE HIGH-PRESSURE LIMIT

In the high-pressure limit, it is safe to assume that the distribution over the levels i is in equilibrium. In this limit the number of collisions that occur on a time scale of 10^{-13} sec is large on the much longer time scale for the unimolecular decay process, which typically occurs on a time scale of 10^{-9} sec. Hence equilibrium over the molecular levels of molecule A can be maintained and the level population n_i replaced by the equilibrium population, i.e., $n_i \sim \exp(-E_i/kT)$. Again since the number of states is large we introduce a density of states in an energy range dE, rather than the individual levels "i". Thus the density of states $\rho(E)$ is defined such that

$$\rho(E)\,dE = \text{number of states in the range } [E;\, E + dE] \qquad (8.14)$$

and hence the equilibrium distribution becomes

$$n_{eq}(E) = \frac{\rho(E)\exp(-E/kT)}{Q} \qquad (8.15)$$

where the partition function is

$$Q = \int_0^\infty dE\, \rho(E)\exp(-E/kT) \qquad (8.16)$$

Furthermore at equilibrium we have, due to the principle of detailed balance,

$$n(E)k(E, E') = n(E')k(E', E) \qquad (8.17)$$

and the expression for the unimolecular rate constant becomes

$$k_{\text{uni}}^\infty = \frac{\int dE\, k(E)\rho(E)\exp(-E/kT)}{Q} \qquad (8.18)$$

If we on the other hand introduce the probability $P(E, T)\,dE$ that molecule A has an energy in the range $[E;\, E + dE]$, where

$$P(E, T) = \frac{\rho(E)}{Q}\exp(-E/kT) \qquad (8.19)$$

(the same expression as n_{eq} in Eq. (8.15)) we obtain

$$k_{\text{uni}}^\infty(T) = \int_0^\infty dE\, k(E)P(E, T) \qquad (8.20)$$

where $k(E)$ remains to be evaluated. As mentioned above, the molecules can react only if they possess an energy over a certain value E_0, taken to be the top

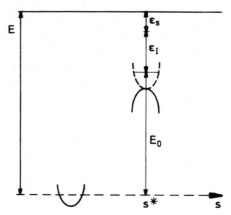

Fig. 8.3. Energetics for unimolecular reactions in the RRKM theory.

of the barrier for decomposition into products (see Fig. 8.3). Thus $k(E) = 0$ for $E < E_0$. For energies above this value we can estimate a microcanonical rate constant $k(E)$ using arguments from microcanonical transition state theory.

In transition state theory one operates with the concept of "the point of no return." This means that if the system has passed over this point along the reaction path, it will not return to reactants but proceed all the way to form products. This point, usually located at the top of the barrier, will be denoted by s^*. Another assumption is that at s^* all quantum states are equally probable (this being a statistical assumption). At the transition state, the excess energy $E - E_0$ can be divided into kinetic energy for the one-dimensional motion along the reaction path, ϵ_s, and internal (rotational and vibrational) energy of the "supermolecule," ϵ_I, i.e.,

$$E - E_0 = \epsilon_s + \epsilon_I \tag{8.21}$$

In order to determine the rate at which the transition state is crossed we must determine the number of states with a given value of kinetic energy ϵ_s along the path $N(\epsilon_s, \epsilon_I)$ and the rate with which these states pass s^*. Assume that one half of the molecules have positive and the other half negative velocity at s^*. Then the number that pass per unit length and time (the rate constant $k^{\#}(\epsilon_s)$ where # indicates the transition state), would be

$$k^{\#}(\epsilon_s) = \frac{\sqrt{2\epsilon_s/m}}{2\delta s} \tag{8.22}$$

where δs is a finite length along the reaction path at the transition state. The precise definition of the mass m is not important since it vanishes later. The microcanonical rate constant $k(E)$ can be obtained by summing over these rates

$$k(E) = \sum_{\epsilon_I = 0}^{E - E_0} k^{\#}(\epsilon_s) \frac{\rho^{\#}(\epsilon_s, \epsilon_I)}{\rho(E)} \qquad (8.23)$$

where the last factor is the probability that, out of the quantum states available for the molecule at the transition state, states are formed with a given part (ϵ_s) of the energy in the reaction path motion. The ratio can also be written as

$$\frac{\rho^{\#}(\epsilon_s, \epsilon_I)}{\rho(E)} = \frac{\rho^{\#}(\epsilon_s) N^{\#}(\epsilon_I)}{\rho(E)} \qquad (8.24)$$

where $\rho^{\#}(\epsilon_s, \epsilon_I)$, $\rho(E)$, and $\rho^{\#}(\epsilon_s)$ denote density of states and $N^{\#}(\epsilon_I)$ the number of vibrational-rotational states having an energy equal to ϵ_I. The density of states for the one-dimensional translational motion can be estimated using the quantum mechanical treatment of the particle in a box of length δs. The energy for a particle in a one-dimensional box is given as

$$\epsilon_s = \frac{n^2 h^2}{8 m \delta s^2} \qquad (8.25)$$

Due to nondegeneracy (in this 1-D case) the number of states with energy up to ϵ_s is

$$n = \frac{\sqrt{8 m \epsilon_s} \delta s}{h} \qquad (8.26)$$

and the density of states $\rho = dn/d\epsilon$ is

$$\rho^{\#}(\epsilon_s) = \sqrt{\frac{2m}{\epsilon_s}} \frac{\delta s}{h} \qquad (8.27)$$

Introducing this in the expression for $k(E)$ we get

$$k(E) = \frac{1}{h\rho(E)} \sum_{\epsilon_I = 0}^{E - E_0} N^{\#}(\epsilon_I) \qquad (8.28)$$

where the summation gives the total number of rotational-vibrational states of the activated complex having an energy less than or equal to $E - E_0$. This number is denoted by $N^{\#}(E - E_0)$. Finally, we obtain the following expression for the unimolecular high-pressure limit of the rate constant

$$k_{uni}^{\infty}(T) = \frac{1}{hQ} \int_{E_0}^{\infty} dE \, N^{\#}(E - E_0) \exp(-E/kT) \qquad (8.29)$$

The integral is equal to $kTQ^{\#} \exp(-E_0/kT)$ where $Q^{\#}$ is the partition function for the internal degrees of freedom of the activated complex, i.e., A^*. Similarly Q is the partition function for the internal degrees of freedom of the nonactivated molecule A.

The final expression is then

$$k_{uni}^{\infty}(T) = \frac{kT}{h} \frac{Q^{\#}}{Q} \exp(-E_0/kT) \qquad (8.30)$$

leading to a transition state expression.

8.2   THE LOW-PRESSURE LIMIT

Let us return to expression (8.13) and consider the low-pressure limit. In this limit, the upper states (i.e., states for which the reaction can occur) react so slowly that we can assume that the rate at which these levels are populated is equal to the unimolecular decay rate, i.e., that

$$[M] \int dE' \, k(E', E)n(E') \sim k(E)n(E) \qquad (8.31)$$

This implies that the unimolecular decay rate is faster than the deactivation rate due to the small level population $n(E)$ for $E > E_0$ and the small number of collisions (due to the low pressure). Thus we get from Eq. (8.13)

$$k_{uni}^{0} = [M] \frac{\int dE \, k_2(E)n(E)}{\int n(E) \, dE} \qquad (8.32)$$

where we have introduced:

$$k_2(E) = \int dE' \, k(E, E') \qquad (8.33)$$

which defines $k_2(E)$ as the total deactivation rate from a level with energy in the range $[E; E + dE]$. Introducing the probability of being in this energy range,

$P(E)\,dE$, where

$$P(E) = \frac{n(E)}{\int dE\, n(E)} \qquad (8.34)$$

we get the following expression for the low-pressure rate constant

$$k^0_{uni} = [M] \int_{E_0}^{\infty} dE\, k_2(E)P(E) \qquad (8.35)$$

The rate constant k_2 is, in general, difficult to determine. An approximate expression, however, can be obtained by assuming that it is equal to the collision number Z^0, i.e., that every collision deactivates the molecule. Therefore

$$k^0_{uni} = [M]Z^0 \int_{E_0}^{\infty} dE\, P(E), \qquad (8.36)$$

where the integral gives the probability of molecules having energies above the energy level E_0, and Z^0, according to collision theory, is

$$Z^0 = \sqrt{\frac{8kT}{\pi\mu}}\, \pi d^2 \qquad (8.37)$$

The reduced mass μ is $m_M m_A/(m_M + m_A)$ and d is the hard-sphere diameter (see Chapter 4) for the collisions between A and the bath molecules M.

8.3   CALCULATION OF THE DENSITY OF STATES

In order to evaluate the microcanonical rate constant, $k(E)$, it is necessary to be able to calculate $\rho(E)$, the density of states, and the number of states, $N^{\#}(E - E_0)$, of an activated complex. Since the number of states in an energy interval $[E;\, E+dE]$ is $N(E+dE)-N(E)$ and the density of states is $[N(E+dE)-N(E)]/dE$ we can see that $\rho(E)$ is given as the derivative of $N(E)$, i.e.,

$$\rho(E) = \frac{dN(E)}{dE}, \qquad (8.38)$$

assuming $N(E)$ is a continuous function of E, this being most often the case.

8.3.1 Rotational Motion

To evaluate the number of states for the rotational degrees of freedom we consider as a model a symmetric top molecule, for which the rotational energy is given by

$$E_{JK} = BJ(J+1) + (A-B)K^2 \tag{8.39}$$

where the rotational constants are given in terms of the moments of inertia with respect to the principal axes. For the prolate symmetric top we have $I_a < I_b = I_c$, $A = \hbar^2/2I_a$, and $B = \hbar^2/2I_b$. Each rotational state is degenerate, the degeneracy factor being $2(2J+1)$, apart from the $K = 0$ level, where it is $2J+1$. The rotational motion can be thought of as consisting of two independent rotations: one about the b- and c-axes (a two-dimensional rotation, $d_i = 2$) with degeneracy $2J + 1$, and a one-dimensional rotation about the a-axis, $d_i = 1$, with degeneracy 2. The degeneracy can then, (if we neglect the 1 in $2J + 1$), be written as $2J_i^{d_i-1}$. If we consider p such rotors we find that the total number of rotational states having energies up to $E_{\rm rot}$ is

$$N(E_{\rm rot}) = \sum_{E=0}^{E_{\rm rot}} \Pi_{i=1}^p 2J_i^{d_i-1} \tag{8.40}$$

where the rotational energy can be written as

$$E_{\rm rot} = \sum_i \frac{\hbar^2}{2I_i} J_i^2 \tag{8.41}$$

and where the moments of inertia are denoted I_i.

If we assume that the rotational states are densely spaced then we can replace the summation with an integral and we obtain

$$N(E_{\rm rot}) = 2^p \int dJ_1 \ldots \int dJ_p J_1^{d_1-1} \ldots J_p^{d_p-1} = E_{\rm rot}^{s/2} \frac{\Pi_{i=1}^p \Gamma\left(\dfrac{di}{2}\right)\left(\dfrac{2I_i}{\hbar^2}\right)^{d_i/2}}{\Gamma\left(1+\dfrac{s}{2}\right)} \tag{8.42}$$

with the constraint

$$\sum_{i=1}^p \frac{\hbar^2}{2I_i} J_i^2 \le E_{\rm rot} \tag{8.43}$$

Thus the total number of rotational states having energies up to E_{rot} is obtained as

$$N(E_{rot}) = \frac{Q_r}{\Gamma\left(1 + \frac{1}{2}s\right)} (E_{rot}/kT)^{s/2} \qquad (8.44)$$

where $s = \sum_i d_i$ and the rotational partition function

$$Q_r = \Pi_{i=1}^p \Gamma\left(\frac{1}{2} d_i\right) \left(\frac{2I_i kT}{\hbar^2}\right)^{d_i/2} \qquad (8.45)$$

has been introduced. The density of states is obtained by simply differentiating (8.45) with respect to the energy,

$$\rho_r(E_{rot}) = \frac{Q_r}{(kT)^{s/2}\Gamma(s/2)} E_{rot}^{(s/2)-1} \qquad (8.46)$$

Note that, although Eqs. (8.44) and (8.46) appear to be temperature dependent, they are, of course, not. At first sight it may seem unnecessary to have such a general expression for the number of rotational states, since any molecule only has three rotational degrees of freedom. However, polyatomic molecules and transition state complexes may have several slightly hindered torsional motions that are better treated as free rotations than as vibrational motions, i.e, p may be larger than 3.

8.3.2 Vibrational Motion

In order to evaluate the number of vibrational states with energies less than or equal to E_{vib} for a molecule with s vibrational degrees of freedom, we need to perform the summation

$$N(E_{vib}) = \sum_{v_1=0} \sum_{v_2=0} \cdots \sum_{v_s=0} 1 \qquad (8.47)$$

subject to the constraint

$$\hbar \sum_{i=1}^{s} \omega_i v_i \leq E_{vib} \qquad (8.48)$$

Using a classical approach (i.e., a continuous v_i) and replacing the summation with an integral we get

$$N(E_{\text{vib}}) = \int_{v_1 = 0} \int_{v_2 = 0} \cdots \int_{v_s = 0} \Pi_i \, dv_i = \frac{E_{\text{vib}}^s}{s! \displaystyle\prod_{i=1}^{s} \hbar\omega_i} \tag{8.49}$$

where E_{vib}^s is the vibrational energy of the s vibration. The density of states is obtained as:

$$\rho_v(E_{\text{vib}}) = \frac{E_{\text{vib}}^{s-1}}{\Gamma(s) \displaystyle\prod_{i=1}^{s} \hbar\omega_i} \tag{8.50}$$

In order to calculate the density of states including both the rotational and vibrational degrees of freedom we use

$$\rho(E) = \int_0^E dE_{\text{rot}} \, \rho_v(E - E_{\text{rot}}) \rho_{\text{rot}}(E_{\text{rot}}) \tag{8.51}$$

where $E = E_{\text{vib}} + E_{\text{rot}}$. If we insert the expressions obtained above for the vibrational and rotational density of states, we get

$$\rho(E) = \frac{1}{\displaystyle\prod_{i=1}^{n} \hbar\omega_i} \frac{Q_r}{(kT)^{r/2}} \frac{E^{n-1+r/2}}{\Gamma(n-1+r/2)} \tag{8.52}$$

where we have introduced n as the number of vibrational modes and r as the dimension of the rotor $r = \sum_i d_i$. For corrections to this expression—corrections due to, for example, the zero point vibrational energy—see the discussions in references [5,6].

REFERENCES

[1] F. A. Lindemann, Trans. Faraday Soc. **17,**598(1922).

[2] G. N. Hinshelwood, Proc. Roy. Soc. London **A113,**230(1926); L. S. Kassel, J. Phys. Chem. **32,**225, 1065(1928); O. K. Rice and H. C. Ramsperger, J. Am. Chem. Soc. **49,**1617(1927); R. A. Marcus and O. K. Rice, J. Phys. Chem. **55,**894(1951).

[3] N. B. Slater, Proc. Cambr. Phil. Soc. **35,**56(1939).

[4] R. A. Marcus, J. Chem. Phys. **20,**359(1952).

[5] R. G. Gilbert and S. C. Smith, "Theory of Unimolecular and Recombination Reactions," Blackwell Scientific, Oxford, 1990.

[6] D. M. Hirst, "A Computational Approach to Chemistry," Blackwell Scientific, Oxford, 1990.

EXERCISES

1. Show that Eq. (8.30) can be obtained from Eq. (8.29).

2. Compare Eq. (8.45) with the partition function given in Table 5.2.

3. Use Eqs. (8.36) and (8.19) to find an expression for k_{uni}^0. Only the vibrational degrees of freedom should be included. Using the incomplete gamma function defined by

$$\Gamma(\alpha, x) = \int_x^\infty dt \exp(-t) t^{(\alpha - 1)} \qquad (8.53)$$

obtain a simplified expression for the rate constant using the expansion

$$\Gamma(\alpha, x) \sim x^{(\alpha - 1)} \exp(-x) \left[1 - \frac{\Gamma(2 - \alpha)}{x\Gamma(1 - \alpha)} \right] \qquad (8.54)$$

Use the frequencies for the reactant (R) and the transition state (T) in Table 8.1 to calculate the unimolecular rate constant in the high-pressure limit at $T = 750$ K for the elimination of HCl from chloroethane. Use

$$k_{uni}^\infty(T) = \frac{kT}{h} \frac{Q_{vib}\#}{Q_{vib}} \exp(-E_0/kT) \qquad (8.55)$$

where the rotational motion is neglected and $E_0 = 228$ kJ/mol including zero point vibrational energy difference.

TABLE 8.1. Vibrational Frequencies (in cm^{-1}) for the Elimination of HCl from Chloroethane

Reactant (R)	Transition State (T)
3014	3100(4)
2986	2200
2977	2000
2946	1450(2)
2887	1300
1470	1150(4)
1448(2)	700(3)
1385	235
1289, 1251, 1081	
1030, 964, 786	
677, 336, 250	

The degeneracies are in parentheses. The reaction path degeneracy is $\sigma = 1$. Data from ref. [5].

Experimentally the unimolecular rate constant has often been expressed in terms of an Arrhenius expression:

$$k^\infty_{uni}(T) = A_\infty \exp(-E_\infty/kT) \qquad (8.56)$$

Use the expression (8.55) derived above to find E_∞ and A_∞.

Calculate A_∞ and E_∞ at 750 K and compare the result with that obtained experimentally in the temperature range 714 K to 767 K:

$$k^\infty_{uni\,(exp)} \sim 10^{13.33\pm0.09}s^{-1} \exp\left(-(235.55\pm1.25\ kJ/mol)/kT\right) \qquad (8.57)$$

4. Show that Eq. (8.52) holds. Use the fact that for large values of n we have

$$\sum_{k=0}^{n-1} \binom{n-1}{k} \frac{(-1)^k}{(k+r/2)} \sim \frac{\Gamma(r/2)\Gamma(n)}{\Gamma(n-1+r/2)} \qquad (8.58)$$

Eq. 8.58 is from the book by E. T. Whittaker and G. N. Watson, "A Course of Modern Analysis," Cambridge University Press, 1927, p. 260.

5. The total available energy of a system, $E = i\hbar\omega$, is shared by s oscillators. According to probability theory, the probability that the system has the energy $E_0 = m\hbar\omega$ in one particular oscillator is

$$\frac{(i-m+s-1)!i!}{(i-m)!(i+s-1)!} \qquad (8.59)$$

Kassel assumed that the rate constant k_a is proportional to this probability. Show that in the limit of large energy ($i \to \infty$) we have:

$$k_a(E) = A\left(1 - \frac{E_0}{E}\right)^{s-1} \qquad (8.60)$$

where A is a constant. Introduce this expression in the Eq. (8.8), introduce the "effective" collision frequency $\omega = [M]k_2$, and use Eq. (8.19) with $\rho(E) = \rho_{vib}(E)$ to obtain an expression for the unimolecular rate constant. Discuss the high- and low-pressure limits. Find the effective activation energy in both limits (assume that ω is independent of temperature). In the Kassel theory s is considered as a parameter, the number of "effective" oscillators participating in the energy distribution process.

For the dissociation of CH_3NNCH_3 by collisions with nitrogen at 600 K, the following pressure dependence was measured:

P (atm)	$k_{uni}(sec^{-1})$
0.0743	0.00213
0.0483	0.00176
0.0213	0.00145
0.00199	0.00069

In the high-pressure limit $k_{uni}^{\infty} = 0.00310$ sec^{-1} and the activation energy $E_0 = 219.4$ kJ/mol. Assume that the number of effective oscillators is $s = 10$ or $s = 15$ and calculate the rate constant at a pressure of 0.002 atm using the hard-sphere model with d = 3 Å.

6. Use the quantum mechanical expression, Eq. (8.59), and the expression (8.61) for the fraction of molecules in quantum state j,

$$P_j = \frac{g_j \exp(-j\hbar\omega/kT)}{\sum_j g_j \exp(-j\hbar\omega/kT)} \qquad (8.61)$$

to derive a quantum mechanical expression for k_{uni}

7. The barrier for the unimolecular reaction:

$$H_2CO \rightarrow H_2CO^{\#} \rightarrow H_2 + CO \qquad (8.62)$$

is 363 kJ/mol. The principal moments of inertia for formaldehyde are $I_a = 1.76$ amu Å^2, $I_b = 12.9$ amu Å^2, and $I_c = 14.7$ amu Å^2. For the molecule in the transition state we have $I_a = 1.93$ amu Å^2, $I_b = 14.0$ amu Å^2, and $I_c = 15.9$ amu Å^2. Calculate using the data in Table 8.2 the rate constant for the unimolecular reaction in the high-pressure limit at 1000 K and 2000 K.

TABLE 8.2. Vibrational Frequencies for Formaldehyde (in cm^{-1})

Frequency	H_2CO	$H_2CO^{\#}$
ω_1	2963	3145
ω_2	1802	1881
ω_3	1559	1359
ω_4	1209	812
ω_5	3038	878
ω_6	1294	i1935

"i" is the imaginary unit.

9

CLASSICAL DYNAMICS

The dynamics of chemical reactions, here specifically the motion of nuclei, can often be well approximated by classical mechanics. For heavy atoms, quantum effects such as tunneling, large vibrational energy spacing, contributions from resonances, and so on, are of little importance. This is especially true if averaged quantities, such as reaction rate constants averaged over internal rotational and/or vibrational states, are calculated. This is fortunate, since the solution of the classical equations of motion usually poses no numerical problem even for a large number of atoms.

Given a potential energy surface, the asymptotic molecular states are initially given specific rotational-vibrational quantum numbers. A classical trajectory is calculated and analyzed asymptotically in order to determine whether the trajectory is reactive or inelastic. In addition, the information about final coordinates and momenta may be used to find the classical quantities (actions) corresponding to the vibrational and rotational quantum numbers. Hence state-to-state probabilities and rates are in principle obtainable even from classical dynamics, such calculations being called quasi-classical.

Let us consider a system consisting of three atoms A, B and C. Atom A approaches the diatomic molecule BC (see Fig. 9.1) and the following coordinates and momenta are used to describe the motion of A with respect to the center of mass of BC:

$$(X\ Y\ Z\ P_X\ P_Y\ P_Z) \tag{9.1}$$

For the motion of the diatomic molecule we use the coordinates and momenta

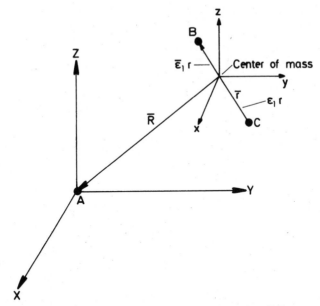

Fig. 9.1. Coordinates used for a classical treatment of A + BC collisions; $\epsilon_1 = m_B/(m_B + m_C)$ and $\bar{\epsilon}_1 = 1 - \epsilon_1$.

$$(x\,y\,z\,p_x\,p_y\,p_z) \tag{9.2}$$

The kinetic energy of the motion of A relative to the center of mass of BC is:

$$E_{\text{kin}} = \frac{1}{2\mu}(P_X^2 + P_Y^2 + P_Z^2) = \frac{P_R^2}{2\mu} + \frac{L^2}{2\mu R^2} \tag{9.3}$$

where the reduced mass μ is

$$\mu = \frac{m_A(m_B + m_C)}{m_A + m_B + m_C} \tag{9.4}$$

L is the orbital angular momentum, and P_R the momentum for the motion along the R-axis. The R-axis being that connecting A and BC. For the motion of the diatomic molecule we have

$$E_{\text{int}} = \frac{1}{2m}(p_x^2 + p_y^2 + p_z^2) + v(r) = \frac{p_r^2}{2m} + \frac{j^2}{2mr^2} + v(r) \tag{9.5}$$

where m is the reduced mass of the diatomic molecule, $m = m_B m_C/(m_B + m_C)$,

j the rotational angular momentum, and $v(r)$ the intramolecular potential, e.g., a Morse potential.

If a reaction occurs it is convenient to describe the system using the $\mathbf{r'R'}$ coordinates shown in Fig. 9.2. They may, however, be easily expressed in terms of the original initial coordinates $\mathbf{r}, \mathbf{R}$:

$$\mathbf{r'} = (x', y', z') = -(X + \epsilon_1 x, Y + \epsilon_1 y, Z + \epsilon_1 z) \tag{9.6}$$
$$\mathbf{R'} = (X', Y', Z') \tag{9.7}$$
$$X' = -X\epsilon_2 + x(1 - \epsilon_2\epsilon_1) \tag{9.8}$$
$$Y' = -Y\epsilon_2 + y(1 - \epsilon_2\epsilon_1) \tag{9.9}$$
$$Z' = -Z\epsilon_2 + z(1 - \epsilon_2\epsilon_1) \tag{9.10}$$

where $\epsilon_1 = m_B/(m_B+m_C)$ and $\epsilon_2 = m_A/(m_A+m_C)$. Once the system is initialized in channel 1 (A + BC), the equations of motion are solved using the Hamiltonian

$$H = \frac{1}{2\mu}(P_X^2 + P_Y^2 + P_Z^2) + \frac{1}{2m}(p_x^2 + p_y^2 + p_z^2) + V(R_{AB}, R_{BC}, R_{AC}). \tag{9.11}$$

Thereby the following equations of motion are obtained

$$\dot{X} = P_X/\mu \tag{9.12}$$
$$\dot{P}_X = -\frac{\partial V}{\partial X} \tag{9.13}$$
$$\dot{x} = p_x/m \tag{9.14}$$
$$\dot{p}_x = -\frac{\partial V}{\partial x} \tag{9.15}$$

along with the corresponding equations for the y- and z-components. After the

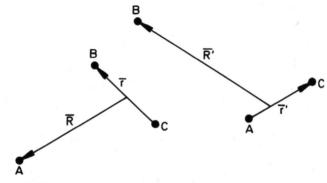

Fig. 9.2. Change in the relative position vectors for a chemical reaction.

collision, the magnitudes of the three distances R_{AB}, R_{BC}, and R_{AC} determine the channel to which the trajectory contributes

$$\text{Channel 1: } R_{AB} \text{ and } R_{AC} \text{ large} \tag{9.16}$$

$$\text{Channel 2: } R_{AB} \text{ and } R_{BC} \text{ large} \tag{9.17}$$

$$\text{Channel 3: } R_{AC} \text{ and } R_{BC} \text{ large} \tag{9.18}$$

The coordinates for channel 2 are indicated above and the corresponding momenta are

$$\frac{p_{x'}}{m'} = \dot{x}' = -\dot{X} - \epsilon_1 \dot{x} \tag{9.19}$$

$$\frac{P_{X'}}{\mu'} = \dot{X}' = -\dot{X}\epsilon_2 + \dot{x}(1 - \epsilon_2\epsilon_1) \tag{9.20}$$

or

$$p_{x'} = -\frac{m'}{\mu} P_X - \epsilon_1 \frac{m'}{m} p_x \tag{9.21}$$

$$P_{X'} = -\frac{\mu'}{\mu} P_X \epsilon_2 + \frac{\mu'}{m} p_x(1 - \epsilon_2\epsilon_1) \tag{9.22}$$

along with corresponding equations for the y- and z-components. The reduced channel masses are $m' = m_A m_C/(m_A + m_C)$ and $\mu' = m_B(m_A + m_C)/(m_A + m_B + m_C)$. Similar equations can be obtained for channel 3.

9.1 INITIALIZATION

In order to compute the classical trajectories we must initialize the six coordinates and momenta. For the motion of atom A relative to the center of mass of BC we can specify the following initial values (see Fig. 9.3)

$$(X, Y, Z) = (0, -\sqrt{R_0^2 - b^2}, b) \tag{9.23}$$

$$(P_X, P_Y, P_Z) = (0, P_Y, 0), \tag{9.24}$$

indicating that the motion is started in the impact parameter plane. The initial value of the momentum P_Y is

$$P_Y = \sqrt{P_0^2 + L^2/R^2} \tag{9.25}$$

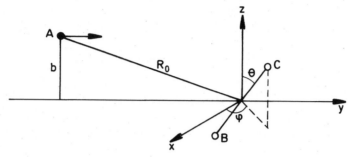

Fig. 9.3. Specification of an initial coordinate system for a classical mechanical treatment of an A + BC collision in the impact parameter picture.

where P_0 is the momentum at $R \rightarrow \infty$ and the orbital angular momentum is $L = P_0 b$. The distance R_0 should be taken to be so large that the interaction potential vanishes at that separation. Then all that takes place over the distance $R = \infty$ to $R = R_0$ is that kinetic energy is converted to orbital energy. For the diatomic molecule we can specify the orientation through the polar angles θ and ϕ,

$$x = r(\kappa) \sin \theta \cos \phi \tag{9.26}$$

$$y = r(\kappa) \sin \theta \sin \phi \tag{9.27}$$

$$z = r(\kappa) \cos \theta \tag{9.28}$$

where θ is chosen at random between 0 and π and ϕ at random between 0 and 2π. The BC distance can also be chosen randomly between the two classical turning points r_- and r_+. For a harmonic oscillator they will be given as:

$$r_\pm = r_e \pm \sqrt{\frac{2\hbar\omega}{k}\left(v + \tfrac{1}{2}\right)} \tag{9.29}$$

where k is the force constant and r_e is the equilibrium bond distance of BC. The solution of the equations of motion for a harmonic oscillator,

$$\dot{p}_r = -\frac{\partial H}{\partial r} = -k(r - r_e) \tag{9.30}$$

$$\dot{r} = \frac{\partial H}{\partial p_r} = \frac{p_r}{m}, \tag{9.31}$$

gives

$$r = r_e + A \cos{(\omega t + \delta)} \tag{9.32}$$

where $A = \sqrt{2\hbar(n + \frac{1}{2})/m\omega}$, and δ is a phase angle that can be chosen randomly. Thus we see that we can introduce a random variable κ between 0 and 1 such that

$$r = r_e + A \cos{(\pi\kappa)} \tag{9.33}$$

So far we have introduced three random variables (θ, ϕ, κ) and have specified 9 out of the 12 initial coordinates and momenta.

Considering the last three variables (the momenta for the motion of the diatomic molecule) we note that the kinetic energy is divided into two parts: one part having to do with the motion along the r-axis (the vibrational motion) and the other being perpendicular to the axis (the rotational motion). Thus

$$\mathbf{p} = \mathbf{p}_r + \mathbf{p}_\perp \tag{9.34}$$

The direction of $\mathbf{p}_r$ is the same as $\mathbf{r}$, i.e.,

$$p_{rx} = p_r \sin\theta \cos\phi \tag{9.35}$$

$$p_{ry} = p_r \sin\theta \sin\phi \tag{9.36}$$

$$p_{rz} = p_r \cos\theta \tag{9.37}$$

The direction of $\mathbf{p}_\perp$ in a plane perpendicular to the r-axis is however unspecified and can be chosen at random (see Fig. 9.4). Thus in a local, body-fixed coordinate system, this orientation can be given by the vector $(\cos\eta, \sin\eta, 0)$.

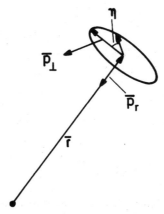

Fig. 9.4. Specification of the random orientation of the vector $\mathbf{p}_\perp$ in a plane perpendicular to the interatomic axis.

In the space-fixed coordinate system, we then get the following expressions

$$p_{\perp x} = \frac{j}{r} (\sin \phi \cos \eta + \cos \theta \cos \phi \sin \eta) \tag{9.38}$$

$$p_{\perp y} = \frac{j}{r} (-\cos \phi \cos \eta + \cos \theta \sin \phi \sin \eta) \tag{9.39}$$

$$p_{\perp z} = -\frac{j}{r} \sin \theta \sin \eta \tag{9.40}$$

where the angle η is chosen between 0 and 2π, and j, the rotational angular momentum, is defined to be $\mathbf{j} = \mathbf{r} \times \mathbf{p}$.

The rate constant for the reaction involving given initial vibrational and rotational states can now be obtained by averaging over a number of trajectories, while varying the initial kinetic energy and impact parameter, i.e.,

$$k_r^{vj}(T) = \sqrt{\frac{8kT}{\pi\mu}} 2\pi \int_0^{b_{max}} b \, db \int_0^\infty d(\beta E_{kin}) \beta E_{kin} \exp(-\beta E_{kin})$$

$$\times \frac{1}{2(2\pi)^2} \int_0^\pi d\theta \sin \theta \tag{9.41}$$

$$\times \int_0^{2\pi} d\phi \int_0^{2\pi} d\eta \int_0^1 d\kappa \, P_r(E_{kin}, b, \theta, \phi, \eta, \kappa; j, v) \tag{9.42}$$

where $P_r = 1$ for a reactive trajectory and $P_r = 0$ for a nonreactive trajectory. Assuming, in addition, a Boltzmann distribution over the initial rotational state, we obtain the "vibrational resolved" rates:

$$k_r^v(T) = \frac{1}{Q_r} \sum_{j=0}^{j_{max}} (2j + 1) \exp(-\beta E_{vj}) k_r^{vj}(T) \tag{9.43}$$

where Q_r is the rotational partition function for the BC molecule.

9.2   FINAL STATE ANALYSIS

Using the information obtained from the final coordinates and momenta it is possible to assign a set of quantum numbers v and j to a specific trajectory.

This is done using

$$\mathbf{j} = \mathbf{r}' \times \mathbf{p}' \qquad (9.44)$$

where the $'$ indicates that the final coordinates and momenta are used. Once the rotational angular momentum is determined we can find the vibrational energy:

$$E_{vib} = E_{int} - E_{rot} \qquad (9.45)$$

From the vibrational energy the vibrational quantum number v is obtained using the expression for the energy levels of a harmonic or a Morse oscillator. At higher values of v this expression is not accurate and the following semiclassical formula, based on the Bohr-Sommerfeld quantization,

$$v = -\frac{1}{2} + \frac{1}{h} \oint p_r \, dr \qquad (9.46)$$

can instead be used. Thus we obtain:

$$v = -\frac{1}{2} + \frac{1}{\pi \hbar} \int_{r_-}^{r_+} dr \sqrt{2m \left(E_{int} - v(r) - \frac{j^2}{2mr^2} \right)} \qquad (9.47)$$

Classical trajectory calculations are straightforward to perform, and oftentimes the predicted rate constants are accurate. However for energies near threshold, the cross sections may be underestimated due to quantum tunneling (see Table 9.1).

TABLE 9.1. Comparison of Classical and Quantum Mechanically Calculated Cross Sections For the Reaction $F + H_2 \rightarrow HF + H$

E_{total}/eV	σ_{clas}	σ_{quant}
0.325	0.39^a	1.509
0.350	2.16	3.102
0.375	3.21	4.033
0.400	4.17	4.506
0.425	4.53	4.780
0.450	5.13	5.123

Cross-sections are given in $\mathring{A}^2$.
The total energy is $E_{total} = E_{kin} + E_{vj}$. The initial vibrational rotational states of hydrogen are $(v,j) = (0, 0)$. The barrier for reaction is about 3.77 kJ/mol. Data from ref. [1].
^{a}3% accuracy obtained by running 5000 trajectories.

REFERENCES

[1] D. Neuhauser, R. S. Judson, R. L. Jaffe, M. Baer and D. J. Kouri, Chem. Phys. Lett. **176,**546(1991).

EXERCISES

1. Derive Eqs. (9.21) and (9.22) for channel 3.

2. Use the Bohr-Sommerfeld expression to find the energy levels of a harmonic oscillator.

3. Show that the initial BC distance for a Morse oscillator should be chosen according to the formula:

$$r = r_e + \frac{1}{\beta} \ln\left[-\frac{b}{2a}\left(1 - \sqrt{1 - \frac{4ac}{b^2}} \cos \pi\kappa \right) \right] \qquad (9.48)$$

where β is the Morse parameter (see Eq. (2.9)) and κ is a random number between 0 and 1. Show that the parameters a, b, and c are given by

$$a = E_{vj} - D_e - A j_r^2 \qquad (9.49)$$
$$b = 2D_e - B j_r^2 \qquad (9.50)$$
$$c = C j_r^2 - D_e \qquad (9.51)$$

where $j_r^2 = \hbar^2 j(j + 1)$ and

$$A = \frac{1}{2mr_e^2}\left[1 - \frac{3}{\beta r_e}\left(1 - \frac{1}{\beta r_e} \right) \right] \qquad (9.52)$$

$$B = \frac{2}{m\beta r_e^3}\left(1 - \frac{3}{2\beta r_e} \right) \qquad (9.53)$$

$$C = \frac{1}{2\beta mr_e^3}\left(1 - \frac{3}{\beta r_e} \right) \qquad (9.54)$$

Use

$$H = E_{vj} = \frac{p_r^2}{2m} + \frac{\hbar^2 j(j + 1)}{2mr^2} + D_e[1 - \exp(-\beta(r - r_e))]^2 \qquad (9.55)$$

10

NONADIABATIC TRANSITIONS

Many chemical reactions involve more than one adiabatic (Born-Oppenheimer) potential energy surface and hence transitions among these surfaces (nonadiabatic transitions) have to be considered. Two different descriptions are used in order to deal with such problems, namely the diabatic or the adiabatic approach. In the diabatic formulation the wavefunction is expanded in "asymptotic" electronic state wave functions. For an A + BC system these states refer to terms of the electronic states at the center of mass distance $R \rightarrow \infty$. In the diabatic formulation we then have (ξ_n and ψ_n are the electronic and nuclear wave functions, respectively):

$$\Psi(q, Q) = \sum_n \xi_n(q, Q_o)\psi_n(Q) \tag{10.1}$$

where q and Q denote electronic and nuclear coordinates, respectively, and Q_o denotes the nuclear configuration that is used for determining the asymptotic electronic wave functions. If we insert this expansion in the Schrödinger equation we obtain a set of coupled equations for the nuclear wave functions:

$$(\hat{T}_{kin} - E)\psi_n(Q) = -\sum_m H_{nm}(Q)\psi_m(Q) \tag{10.2}$$

where $\hat{T}_{kin}$ is the nuclear kinetic energy operator and

$$H_{nm} = \langle \xi_n | \hat{H}_{el} | \xi_m \rangle \tag{10.3}$$

the diabatic coupling elements are obtained by integrating over the electronic coordinates. $\hat{H}_{el}$ is the electronic Hamiltonian.

In the adiabatic case, (ξ_n and ϕ_n are the electronic and nuclear wave functions, respectively), one solves the electronic problem for a given nuclear configuration, hence the adiabatic wavefunctions depend parametrically on the nuclear coordinates, i.e.,

$$\hat{H}_{el}(q, Q) \xi_n(q; Q) = W_n(Q) \xi_n(q; Q) \tag{10.4}$$

The wavefunction for the total system can be expanded as

$$\Psi(q, Q) = \sum_n \xi_n(q; Q) \phi_n(Q) \tag{10.5}$$

Inserting this expansion in the Schrödinger equation, we obtain

$$[\hat{T}_{kin}^{el} + V_{ee}(q) + \hat{T}_{kin} + V_{en}(q, Q) + V_{nn}(Q)] \Psi(q, Q) = E\Psi(q, Q) \tag{10.6}$$

where $\hat{T}_{kin}^{el}$ is the electronic kinetic energy operator, V_{ee} and $V_{nn}(Q)$ are the electron-electron and nucleus-nucleus repulsion operators, respectively, and $V_{en}(q, Q)$ is the electron-nucleus attraction operator. We obtain

$$[\hat{T}_{kin} + W_n(Q) - E]\phi_n(Q)$$
$$= \sum_m \left(\sum_i \frac{\hbar^2}{m_i} \left\langle \xi_n \left| \frac{\partial}{\partial Q_i} \right| \xi_m \right\rangle \frac{\partial}{\partial Q_i} + \frac{\hbar^2}{2m_i} \left\langle \xi_n \left| \frac{\partial^2}{\partial Q_i^2} \right| \xi_m \right\rangle \right) \phi_m(Q)$$
$$\tag{10.7}$$

where m_i are the masses of the nuclei and where the right-hand side contains the nonadiabatic coupling terms arising from the nuclear kinetic energy term. Note that the difference between the diabatic and the adiabatic schemes is that in the first case we have nuclear potential, while in the latter there is nuclear kinetic coupling between the two surfaces.

10.1   THE TWO-STATE CASE

In order to proceed, we consider the two-state case (two surfaces) and just a single internuclear distance R. In the diabatic formulation we then have

$$\begin{bmatrix} \dfrac{\hbar^2}{2m}\dfrac{d^2}{dR^2} + E - H_{11}(R) & -H_{12}(R) \\[2ex] -H_{12}(R) & \dfrac{\hbar^2}{2m}\dfrac{d^2}{dR^2} + E - H_{22}(R) \end{bmatrix} \begin{bmatrix} \psi_1(R) \\[1ex] \psi_2(R) \end{bmatrix} = \begin{bmatrix} 0 \\[1ex] 0 \end{bmatrix}$$

$$(10.8)$$

This set of equations can now be solved analytically, providing various simplifications concerning the coupling and the diabatic potential energy surfaces are made. The simplest nontrivial solution is obtained if one assumes that the potentials can be approximated by a simple linear expression, i.e.,

$$H_{ii}(R) = -F_i(R - R_c) \tag{10.9}$$

where R_c is the crossing point and F_i the slope. If the off-diagonal coupling term H_{12} is taken to be constant, the Landau-Zener [1,2] model emerges. The equations can then be solved exactly, using either a JWKB (Jeffreys, Wentzel, Kramers, Brillouin) [3] or a semiclassical "classical path" model. In the case of the latter, one treats the nuclear motion classically, i.e., we introduce

$$E \rightarrow i\hbar \frac{\partial}{\partial t}$$

$$\frac{\hbar}{i}\frac{\partial}{\partial R} = P_R(t),$$

and

$$R(t) = R_c + \frac{P_R}{m}(t - t_c) \tag{10.10}$$

Thus we obtain the following classical path equations:

$$i\hbar \frac{d}{dt}\begin{bmatrix} \phi_1(t) \\ \phi_2(t) \end{bmatrix} = \begin{bmatrix} 0 & H_{12}\exp(i\Delta(t)) \\ H_{21}\exp(-i\Delta(t)) & 0 \end{bmatrix}\begin{bmatrix} \phi_1(t) \\ \phi_2(t) \end{bmatrix} \tag{10.11}$$

where

$$\Delta(t) = \frac{1}{\hbar}\int dt\,(H_{11} - H_{22})$$

and where we have introduced

$$\phi_i = \psi_i \exp\left[\frac{i}{\hbar}\int dt\,(H_{ii} + P_R^2/2m)\right]$$

In this model the transition probability for a transition from the diabatic surface 1 to surface 2 is approximately given as:

$$P_{12}^d = 1 - \exp\left[-2\pi H_{12}(R_c)^2/\hbar v|F_1 - F_2|\right] \tag{10.12}$$

where v is the velocity with which the surface crossing point is passed, i.e., $v = P_R/m$. Thus the probability of staying on the diabatic surface 1, is $1 - P_{12}^d$. Utilizing the adiabatic representation, the probability of going from adiabatic surface 1 to 2 is $P_{12}^a = 1 - P_{12}^d$. The probability for remaining on surface 1 is then $1 - P_{12}^a$, and the probability of going to the potential curve 2 and remaining after the system encounters a turning point to the left of the crossing point (see Fig. 10.1) is $2P_{12}^{a,d}(1 - P_{12}^{a,d})$.

Considering the adiabatic curves in the two-state case we have:

$$W_{ii}(R) = \tfrac{1}{2}(H_{11} + H_{22}) \pm \tfrac{1}{2}\sqrt{(H_{11} - H_{22})^2 + 4H_{12}^2} \tag{10.13}$$

where the upper sign is for $(i = 2)$ and the lower for $(i = 1)$. Thus at the crossing point $R = R_c$ we have $W_{22} - W_{11} = 2H_{12}$. In the adiabatic picture the two

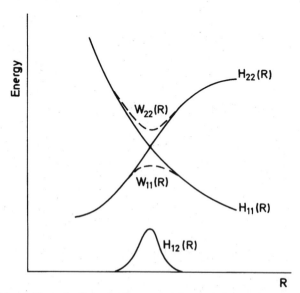

Fig. 10.1. Diabatic H_{ii} and adiabatic W_{ii} potential energy curves for a two-state system.

curves exhibit an avoided crossing at R_c and a classical trajectory $R(t)$ starting at the lower surface will, according to the Landau-Zener expression, have the largest probability for jumping to the upper surface at the avoided crossing. If the system is allowed to jump in the classical path method, the trajectory continues at the upper surface with a kinetic energy $(1/2m)P_R(t_c)^2 - 2H_{12}$. Within the classical path framework the Landau-Zener theory is used in three-dimensional calculations in the so-called trajectory surface hopping method, where classical trajectories are integrated on the adiabatic surfaces, and a discontinuous jump at the avoided crossing is allowed, with a probability obtained using the Landau-Zener formula [4]. For recent corrections to the simple Landau-Zener expression see ref. [5].

REFERENCES

[1] L. D. Landau, Z. Phys. Sov. **2**,46(1932).

[2] C. Zener, Proc. R. Soc. London **A137**,696(1932).

[3] H. Jeffreys, Proc. London Math. Soc. **23**,428(1925); G. Wentzel, Z. Phys. **38**,518(1926); H. A. Kramers, Z. Phys. **39**,828(1926); L. Brillouin, CR Acad. Sci., Paris, **183**,24(1926); N. Fröman and P. O. Fröman, "JWKB Approximation, Contributions to the Theory," North-Holland, Amsterdam, 1965.

[4] J. C. Tully and R. K. Preston, J. Chem. Phys. **55**,562(1971); J. R. Stine and J. T. Muckerman, J. Chem. Phys. **65**,3975(1976).

[5] C. Zhu and H. Nakamura, J. Chem. Phys. **101**,10630(1994), ibid **102**,7448(1995).

EXERCISES

1. Verify Eq. (10.12)

2. The reaction between a metal atom M and a halogen molecule X_2:

$$M + X_2 \rightarrow MX + X \qquad (10.14)$$

occurs through the so-called harpoon mechanism, where an electron at some distance is transferred from the metal to the halogen molecule to form $M^+ + X_2^-$. Subsequently the following reaction occurs

$$M^+ + X_2^- \rightarrow MX + X \qquad (10.15)$$

Find an expression for the reaction probability assuming that it is equal to the Landau-Zener probability and that:

$$H_{11} = -C_6/R^6 \tag{10.16}$$
$$H_{22} = \Delta E - e^2/R \tag{10.17}$$

where ΔE is the difference between the ionization potential I of the atom and the electron affinity A of the halogen molecule. Calculate the Landau-Zener probability at $E = 1$ eV for a model system with $I - A = 2.0$ eV, $H_{12}(R_c) = 0.01$ eV, and a reduced mass of 50 amu.

3. The reaction

$$K + I \rightarrow K^+ + I^-$$

occurs through the harpoon mechanism. The two diabatic potential surfaces near the crossing point R_c are given by $H_{11}(R) \sim C_6/R^6$ and $H_{22}(R) = I - A - e^2/R$, where $I - A = 1.28$ eV. The probability for a nonadiabatic transition at R_c obeys the Landau-Zener expression $P_{12} = \exp(-v_c/v)$.

a. Find an expression for v_c if $H_{11}(R_c) \sim 0$.

b. For the above reaction, we have $|W_{22} - W_{11}| = 2H_{12}(R_c) = 0.025$ eV. Calculate R_c and v_c.

c. What is the probability of forming $K^+ + I^-$ if the turning point effect is included.

d. Find the value of v_c for which this probability has its maximum.

e. Find the reaction probability for $E_{kin} = 1.5$ eV.

11

SURFACE KINETICS

11.1 INTRODUCTION

The interaction of atoms and molecules with a solid surface gives rise to important additional phenomena compared to interactions in the gas phase. For example, the periodicity of the solid gives rise to diffraction patterns, and the solid acts as energy donor or acceptor according to the specific surface temperature. Furthermore some solids (metals) often act as good catalysts, i.e., they lower the activation barrier for the chemical reaction and hence enhance the reaction rate significantly. The precise mechanism for this catalytic effect is not necessarily known in detail. However, it can often be explained phenomologically as an electron transfer from the metal to the incoming molecule. Since the interaction between the electrons of the molecule and the Fermi electrons of the metal tends to change (increase) the electron affinity of the molecule, the transfer can occur at some distance (z) from the surface, even though the energy needed to remove an electron from the solid (the work function ϕ_e) can be as large as 2–3 eV (see Fig. 11.1). This electron donation to the antibonding orbitals of the molecule will weaken the binding energy between the atoms and hence lower the effective dissociation energy. The mechanism is thus similar to the harpoon mechanism for alkali reactions in the gas phase. This specific mechanism is important, since many chemical reactions at surfaces occur with the catalytic dissociation of a diatomic molecule as the rate-determining step.

The dynamics of molecule-surface interactions is complicated by the fact that the activation barrier for dissociative chemisorption depends strongly on the surface site, that the coupling to the surface phonons is strong (except for light atoms or molecules), and that other excitation mechanisms (electron-hole

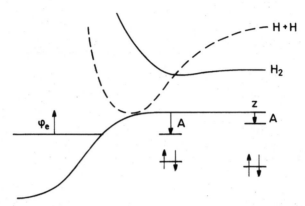

Fig. 11.1. Electron transfer from the metal to an antibonding orbital in the molecule diminishes the dissociation energy. The metal work function is ϕ_e and the electron affinity A.

pair excitation and other collective motion) can occur. Therefore, a realistic description of such phenomena can only be obtained through numerical simulations. Furthermore crystal surfaces are usually not absolutely flat regular arrays of atoms, but rather exhibit anomalies such as steps, kinks, vacancies, and adsorbed atoms. Important dynamical phenomena at the surface involve, aside from the elastic diffractive scattering, inelastic scattering and surface diffusion. Through these processes a molecule can end up as a surface-bound molecule in a physisorbed (weakly bound precursor) state, from which chemisorption into a dissociated state can take place through tunneling through the activation barrier. Another possibility is that the molecule can directly enter the dissociated state by losing a sufficient amount of its kinetic energy to the surface phonons. In any case, the activation barrier for this process is much smaller than for the gas phase dissociation barrier (see Fig. 11.2).

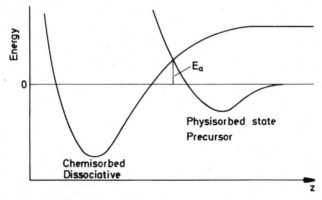

Fig. 11.2. Energetics for a dissociative chemisorption process as a function of the distance (z) from the surface. E_a is the activation barrier.

11.2 WALL COLLISIONS

The mean velocity of molecules approaching a surface (situated in the xy-plane) can be obtained simply by assuming that half of the molecules have a velocity in the negative z-direction (towards the surface), i.e.,

$$\langle v_z \rangle = \sqrt{\frac{2m}{\pi k T}} \int_0^\infty v_z \exp \left(-\frac{1}{2} m v_z^2 / k T \right) dv_z = \sqrt{\frac{2kT}{\pi m}} = \frac{1}{2} \langle v \rangle \qquad (11.1)$$

where m is the mass of the molecule and $\langle v \rangle$, the average speed $\langle v \rangle = \sqrt{8kT/\pi m}$. The number of molecules hitting a unit area per time unit is then

$$Z_s = \frac{1}{2} n \langle v_z \rangle = \frac{1}{4} n \langle v \rangle \qquad (11.2)$$

where n is the number of molecules per unit volume. The pressure, i.e., the force per unit area at the walls of a container is

$$p = nm \langle v_z^2 \rangle = \frac{1}{3} nm \langle v^2 \rangle \qquad (11.3)$$

11.3 THE ADSORPTION ISOTHERM

An important property of a solid surface is its ability to adsorb a given quantity of gas phase molecules. A molecule hitting a clean surface will be adsorbed with a different probability than a molecule hitting a layer of previously adsorbed molecules. Denoting the probability of adsorption on a clean surface by p_0 we find, using Eq. (11.2), that the rate of adsorption is given as

$$v_{ads} = p_0 \frac{n}{4} \langle v \rangle \alpha_0 \qquad (11.4)$$

where α_0 is the fraction of the surface area that is clean. The desorption rate from the part of the surface with just one layer of adsorbed molecules can be expressed as

$$v_{des} = k_1 \sigma_m \alpha_1 \qquad (11.5)$$

where k_1 is the desorption rate constant from this layer, σ_m is the number density of adsorbed molecules (number of molecules per unit area), and α_1 is the fraction of the surface covered by a single layer of adsorbed molecules (Fig. 11.3). At equilibrium we have

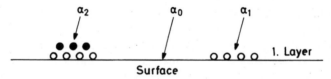

Fig. 11.3. First and second layer gas-surface adsorption.

$$p_0 . \frac{\langle v \rangle}{4} n\alpha_0 = k_1 \sigma_m \alpha_1 \qquad (11.6)$$

For the part of the surface with two adsorbed layers we can write a similar equilibrium condition:

$$p_1 \frac{\langle v \rangle}{4} n\alpha_1 = k_2 \sigma_m \alpha_2 \qquad (11.7)$$

where the probability for adsorption to the first layer atoms p_1 is different from p_0 (as mentioned previously). For layer number N we have

$$p_N \frac{\langle v \rangle}{4} n\alpha_N = k_{N+1} \sigma_m \alpha_{N+1}. \qquad (11.8)$$

For $N \to \infty$, Eq. (11.8) expresses an equilibrium condition between a liquid and a saturated gas phase vapor of the molecules in question.

Hence the value of the number density n should be the saturation value n_s. Furthermore we assume that the adsorption on a multilayer surface is independent of the number of layers, i.e., that

$$\tilde{p} = p_1 = p_2 = \ldots p_\infty \neq p_0 \qquad (11.9)$$

$$k = k_2 = \ldots k_\infty \neq k_1 \qquad (11.10)$$

Thus we get

$$\tilde{p} \frac{\langle v \rangle}{4} n\alpha_1 = k\sigma_m \alpha_2 \qquad (11.11)$$

$$\tilde{p} \frac{\langle v \rangle}{4} n_s = k\sigma_m; \qquad (11.12)$$

or by division of these equations

$$\frac{n}{n_s} = \frac{P}{P_0} = \frac{\alpha_2}{\alpha_1}, \qquad (11.13)$$

where P is the gas phase pressure and P_0 the pressure of the saturated gas vapor. Introducing the quantity $x = P/P_0$ we then have

$$\alpha_2 = x\alpha_1, \tag{11.14}$$

and by introducing this in the subsequent equations for $N > 1$ we get

$$\alpha_i = x^{i-1}\alpha_1 \tag{11.15}$$

Using Eqs. (11.6) and (11.12) we get

$$\alpha_i = Cx^i\alpha_0 \tag{11.16}$$

where $C = p_0 k/\tilde{p}k_1$. Introducing the number density of adsorbed molecules in all layers as

$$\sigma = \sigma_m \sum_{i=1}^{\infty} i\alpha_i \tag{11.17}$$

and using

$$\sum_{i=0}^{\infty} \alpha_i = 1 \tag{11.18}$$

we can derive the expression

$$\frac{x}{\sigma(1-x)} = \frac{1}{C\sigma_m} + x\,\frac{C-1}{C\sigma_m} \tag{11.19}$$

Note that the left-hand side plotted as a function of x is a straight line, and that the slope and interception with the y-axis gives information on C and σ_m, the number density of molecules forming a monolayer.

Introducing $\sigma/\sigma_m = V_{ads}/V_m$ and $x = P/P_0$, we get the adsorption isotherm

$$\frac{P}{V_{ads}(P_0-P)} = \frac{1}{CV_m} + \frac{C-1}{CV_m}\,\frac{P}{P_0} \tag{11.20}$$

The adsorption isotherm is obtained by measuring the amount of adsorbed gas (i.e., V_{ads}) as a function of pressure P. Eq. (11.20) shows that we obtain a straight line if $P/V_{ads}(P_0-P)$ is plotted against P/P_0 with slope $(C-1)/(CV_m)$ (Fig. 11.4). Extrapolation to $P/P_0 = 0$ yields $1/CV_m$. Thus we obtain informa-

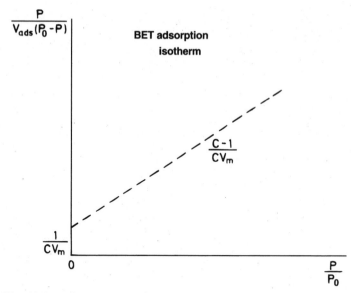

Fig. 11.4. BET (Brunauer, Emmett, Teller) adsorption isotherm [1].

tion on V_m, the amount of gas adsorbed as a monolayer, and on the constant $C = p_0 k / \tilde{p} k_1$, the ratio of adsorption probability for a clean surface and a surface with one or more layers $p_0 / \tilde{p}$ multiplied with the ratio of desorption rates k / k_1.

11.4 SURFACE DIFFUSION

Surface diffusion takes place through hops between potential energy wells (see Fig. 11.5). If the distance between the wells is a and the jump frequency ν the diffusion velocity is

$$v_d = a\nu \tag{11.21}$$

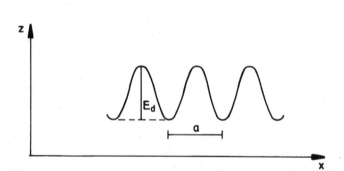

Fig. 11.5. Energy profile for a one-dimensional surface diffusion process.

Experiments show that the diffusion is an activated process, where the molecule surpasses a barrier E_d, and that the jump frequency ν depends upon temperature in an Arrhenius fashion, i.e.,

$$\nu = \nu_0 \exp\left(-E_d/kT_s\right) \tag{11.22}$$

where T_s is the surface temperature. From the random walk expression for the diffusion coefficient (in 1-D) (see also Chapter 12) we then get

$$D(T) = \frac{a^2}{2\tau} = \frac{a^2}{2}\,\nu_0 \exp\left(-E_d/kT_s\right) \tag{11.23}$$

where τ is the time between the jumps. The above expression is in agreement with the result from transition state theory. Considering one-dimensional motion we obtain the rate constant

$$k(T) = \frac{kT}{h}\frac{1}{Q_{\text{vib}}} \exp\left(-E_d/kT_s\right) \tag{11.24}$$

where Q_{vib} is the vibrational partition function for a particle in the potential well. In the high-temperature limit $Q_{\text{vib}} = kT_s/\hbar\omega$, i.e.,

$$k(T) = \frac{\omega}{2\pi} \exp\left(-E_d/kT_s\right) \tag{11.25}$$

Comparing with Eq. (11.23) we have $D(T) = a^2 k(T)/2$. In reality, the molecule can vibrate in three dimensions (x, y, z) in the potential well, and in two dimensions (y, z) at the transition state. We then get

$$k(T) = \frac{\omega_x \omega_y \omega_z}{\omega_y^{\#}\omega_z^{\#}2\pi} \exp\left(-E_d/kT_s\right) \tag{11.26}$$

Since diffusion can occur in both directions along the x-axis the rate constant should be multiplied by a factor of 2.

The diffusion process allows the atoms to move to sites where chemical recombination processes can occur. Since the chemical reaction releases the binding energy of the formed molecule, there is a possibility for overcoming the barrier for desorption. Thus part of the released binding energy will end up as internal energy of the desorbing molecule, part will be used in excitation of the surface phonons and electrons, and the remainder will add to the translational energy of the molecule, which in turn can be sufficient to overcome the activation barrier for desorption. (The surface binding energy of molecules is often weaker than that of atoms.)

The surface reactions between two adsorbed species are known as Langmuir-Hinshelwood processes. Here the reaction rate will be proportional to the coverage (θ_a and θ_b) of the species a and b, i.e., to $k\theta_a\theta_b$. If the reaction occurs directly between an adsorbed and a gas phase species then the process is of the Eley-Rideal type and the rate is proportional to the coverage of the adsorbed species and the partial pressure of the gas phase species. The diffusion constant can be determined experimentally since the process follows Fick's second law (see next chapter).

REFERENCES

[1] S. Brunauer, P. H. Emmett and E. Teller, J. Am. Chem. Soc. **60**,309(1938).

SUGGESTED READING

M. Boudart and G. Djéga-Mariadassou, "Kinetics of Heterogeneous Catalytic Reactions," Princeton University Press, Princeton, 1984.

G. A. Somorjai, "Introduction to Surface Chemistry and Catalysis," John Wiley & Sons, New York, 1994.

EXERCISES

1. Derive Eq. (11.3).

2. Derive Eq. (11.19). Use $\sum_{i=1}^{\infty} x^i = x/(1-x)$ and $\sum_{i=1}^{\infty} ix^i = x/(1-x)^2$.

3. Calculate V_m and σ_m from the following adsorption data for 1 m^2 of surface material: V = 0.10 cm^3, P = 1 torr; V = 0.34 cm^3, P = 150 torr; V = 0.53 cm^3, P = 300 torr; V = 0.80 cm^3, P = 456 torr. The volumes refer to STP conditions and P_0 = 760 torr at 77.3 K.

4. Calculate the diffusion constant at 300 K for sodium on a Cu(001) surface. The lattice constant is a = 2.56 Å; $\hbar\omega$ = 18.5 meV (z-direction) and 5.9 meV (x- and y-direction); $\hbar\omega^{\#}$ = 14 meV (z-direction) and 3.5 meV (y-direction). The barrier height, E_d = 62 meV. Compare the calculated diffusion constant with the experimental value where $D(T) = D_0 \exp(-E_d/kT_s)$ and where D_0 = 1.86 ± 0.04 · 10^{-4} cm^2/sec.

12

CHEMICAL REACTIONS IN SOLUTION

Today, chemical reaction in solution is a significant research area where presently a great deal of progress is being made. Solution is the medium in which most chemical reactions occur. Reactions between molecular species in solution, solutes, will be the subject of interest in the next chapters. This broad category includes reactions between small or moderately sized biological systems. It excludes, however, polymeric, colloidal systems. We will consider diffusion and transport in solution, chemical reactions in solution, and charge transfer processes in solution. Our objective is to provide basic principles about the subject and illustrate some of the basic concepts necessary for understanding chemical reactions in solution (CRIS). We will mainly focus on analytical approaches but will in addition touch on the computational and molecular approaches in this area.

12.1 TRANSPORTATION OF MATTER IN SOLUTION

12.1.1 Mass Transfer

Movement of material from one location in solution to another is due to (i) a difference in electrical or chemical potential at the two locations; (ii) movement of a volume element of solution. Such mass transfer can be classified as migration, diffusion, and convection. Migration is the movement of a charged body under the influence of an electric field (a gradient of the electrical potential). Diffusion is the movement of a species under the influence of a gradient of the chemical potential (a concentration gradient). Convection arises from stirring or hydrodynamic transport. Generally fluid flow occurs because of natural convection (convection caused by density gradients) and forced convection.

Let us consider a certain species j at two points r and s in solution (r and s being an infinitesimal distance from one another). The chemical potentials for species j at the two different positions are not identical; $\mu_j(r) \neq \mu_j(s)$. This difference in the chemical potential corresponds to the work, dw, required to move the material from $r = x$ to $s = x + dx$

$$dw = \mu_j(x + dx) - \mu_j(x)$$

$$= \mu_j(x) + \left[\frac{d\mu_j(x)}{dx} \right] dx - \mu_j(x) = \left[\frac{d\mu_j(x)}{dx} \right] dx \qquad (12.1)$$

at constant pressure and temperature. In classical mechanics the work required to move an object through a distance dx against a force F is

$$dw = -F\,dx \qquad (12.2)$$

The slope of the chemical potential is an effective force, not a real physical force pushing the particles down the slope of the chemical potential. It is a consequence of the second law of thermodynamics and the quest for maximum entropy.

We can express the chemical potential in terms of the activity of the particles. Writing the activity of the particles as a, the chemical potential is given as

$$\mu = \mu^o + RT\ln a \qquad (12.3)$$

where μ^o is the standard chemical potential, R is the gas constant, and T the temperature. If the solution is not uniform, the activity a depends on the position x. From Eqs. (12.2) and (12.3) we obtain

$$F = -\frac{d}{dx}(\mu^o + RT\ln a) = -RT\left(\frac{d\ln a}{dx} \right)_{PT} \qquad (12.4)$$

If the solution is ideal, the activity a may be replaced by the concentration, C, and we have

$$F = -\frac{RT}{C}\left(\frac{dC}{dx} \right)_{PT} \qquad (12.5)$$

We see from Eq. (12.5) that the thermodynamic force F acts in the direction of decreasing concentration.

In what follows we will consider Fick's first law of diffusion from the point of view of thermodynamics. A difference in μ_j over the distance from r to s can arise because of differences in the concentration of the species j at r and s, or from differences in the electric potential at r and s. In general, there will be a flux of species j to even out this difference in μ_j. The flux, J_j, is proportional to the gradient of μ_j as in Eq. (12.6):

$$J_j = -\frac{C_j D_j}{RT} \nabla \mu_j \qquad (12.6)$$

The minus sign arises because the direction of the flux opposes the direction of increasing μ_j. The electrochemical potentials at r and s are given as

$$\mu_j(r) = \mu_j^o + RT \ln a_j(r) + z_j F\phi(r)$$
$$\mu_j(s) = \mu_j^o + RT \ln a_j(s) + z_j F\phi(s) \qquad (12.7)$$

where z_j, F, and $\phi(x)$ are the charge on the species j, Faraday's constant, and the electric potential at position x, respectively. If an element of the solution moves away from s with a velocity v, we have an additional term in the flux equation

$$J_j = -\frac{C_j D_j}{RT} \nabla \mu_j + C_j v \qquad (12.8)$$

Assuming an ideal solution, we obtain

$$J_j = -\frac{C_j D_j}{RT} [\nabla(RT \ln C_j) + z_j F\nabla\phi(r)] + C_j v \qquad (12.9)$$

We will now consider the case where convection is absent, e.g., an unstirred or stagnant solution having no density gradients, thus giving a solution velocity equal to zero. We then have

$$J_j = -C_j D_j \nabla(\ln C_j) - \frac{z_j F}{RT} C_j D_j \nabla\phi(r)$$
$$= -D_j \nabla(C_j) - \frac{z_j F}{RT} C_j D_j \nabla\phi(r) \qquad (12.10)$$

The first term is referred to as the diffusion term and the second is the migration term. If the species j is charged, the flux J_j is equivalent to a current density. Let us consider a linear mass flow system with a cross-sectional area A normal to the axis of mass flow. Here

$$J_j = -\frac{i_j}{z_j FA} \qquad (12.11)$$

where i_j is the current due to the flow of the charged species j. We can define two current components,

$$-J_j = \frac{i_j}{z_j FA} = \frac{i_{j,d}}{z_j FA} + \frac{i_{j,m}}{z_j FA} \qquad (12.12)$$

where $i_{j,d}$ and $i_{j,m}$ are the current components resulting from diffusion and migration, respectively.

12.1.2 Microscopic View of Diffusion

Diffusion, which normally leads to the homogenization of a mixture, occurs by a "random walk" process. A simple picture of this process can be obtained by considering a one-dimensional random walk. Consider a molecule constrained to a one-dimensional path, moving under the influence of the pushing and knocking of solvent molecules. The molecule moves in steps of length λ and one step is made per unit time τ. We can then ask ourselves, "Where will the molecule be after a time t?" We can only answer this by giving the probability of the molecule being at different locations at some time $m\tau$ (see Fig. 12.1). At $t = 0$ the molecule is at $x = 0$; at $t = \tau$ it can either be at $x = -\lambda$ or $x = \lambda$; at $t = 2\tau$ it can be at $x = -2\lambda$, $x = 0$, or $x = 2\lambda$; at $t = 3\tau$ it can be at $x = -3\lambda$, $x = -\lambda$, $x = \lambda$, or $x = 3\lambda$; at $t = 4\tau$ it can be at $x = -4\lambda$, $x = -2\lambda$, $x = 0$, or $x = 2\lambda$, $x = 4\lambda$. At $t = 4\tau$ the probability for the molecule being at $x = 0$ is 6/16; at $x = \pm 2\lambda$ is 4/16; at $\pm 4\lambda$ is 1/16. The general expression for the probability $P(m, r)$ that the molecule is at a given location after m time units ($m = t/\tau$) is given by the binomial coefficient (see ref.[1])

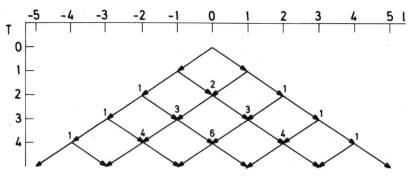

Fig. 12.1. The random walk process.

$$P(m, r) = m![r!(m - r)!]^{-1}(1/2)^m \tag{12.13}$$

where the set of locations is defined by $x = (-m + 2r)\lambda$ with $r = 0, 1, 2, \ldots, m$. The mean (or average) squared displacement of the molecule is

$$\langle \Delta^2 \rangle = \sum_r x^2 P(m, r), \tag{12.14}$$

and for our one-dimensional case we find

$$\langle \Delta^2 \rangle = m\lambda^2 = t/\tau\lambda^2 = 2Dt, \tag{12.15}$$

where D is defined as the diffusion coefficient. The root-mean-squared displacement at time t is thus

$$\sqrt{\langle \Delta^2 \rangle} = \sqrt{2Dt} \tag{12.16}$$

As m approaches infinity we obtain a continuous form for the binomial function: consider N_o molecules located at the origin at $t = 0$. Suppose we wish to find the distribution at a later time t; the result is (see ref.[1])

$$\frac{N(x, t)}{N_o} = dx \frac{1}{\sqrt{4\pi Dt}} \exp\left(-\frac{x^2}{4Dt} \right) \tag{12.17}$$

where $N(x, t)$ is the number of molecules in a segment of width dx centered at a position x.

12.1.3   Fick's Laws of Diffusion

Fick's laws are differential equations describing the flux of a substance and its concentration as functions of time and position. Let us consider the case of linear or one-dimensional diffusion: the flux of a substance j at a given distance x and at a time t, $J_j(x, t)$, is the net mass transfer rate of j in units of mass per unit time per unit area. The flux, $J_j(x, t)$, is the number of moles of j that pass a given location per second per cm^2 of area normal to the axis of diffusion. Fick's first law states that the flux is proportional to the concentration gradient

$$J_j = -D_j \frac{\partial C_j}{\partial x} \tag{12.18}$$

Consider a location x where we assume that we have $N_p(x, t)$ molecules to the right of x and $N_p(x + dx, t)$ molecules to the left of x at time t. During the time increment dt the random walk (i.e., diffusion process) moves half of these molecules dx in either one direction or the other, so the net flux through an area A located at x is given by the difference between the number of molecules moving from left to right and those moving from right to left (dropping the index j) (see Fig. 12.2)

$$-J(x, t) = -\frac{1/2[N_p(x, t)] - 1/2[N_p(x + dx, t)]}{A\, dt} \tag{12.19}$$

Multiplication by $(dx)^2/(dx)^2$ gives

$$-J(x, t) = \frac{C(x + dx) - C(x)}{dx} \frac{(dx)^2}{2\, dt} \tag{12.20}$$

where C is the concentration and is given as

$$C = \frac{N_p}{A\, dx} \tag{12.21}$$

Introducing the diffusion coefficient

$$D = \frac{(dx)^2}{2\, dt} \tag{12.22}$$

leads to the following for Eq. (12.20):

$$-J(x, t) = D\frac{[C(x + dx) - C(x)]}{dx} = D\frac{\partial C}{\partial x} \tag{12.23}$$

i.e., Fick's first law.

Fick's second law concerns the change in concentration with time

$$\frac{\partial C}{\partial t} = D\frac{\partial^2 C}{\partial x^2} \tag{12.24}$$

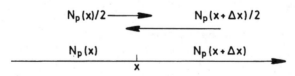

Fig. 12.2. The change in concentration at a location x.

Fig. 12.3. The net flux through an area.

The change in concentration at a location x is given by the difference in the flux into $J(x,t)$ and the flux out of $J(x+dx,t)$ an element of width dx (see Fig. 12.3),

$$\frac{\partial C}{\partial t} = \frac{J(x,t) - J(x+dx,t)}{dx} \tag{12.25}$$

A Taylor expansion of $J(x+dx,t)$ leads to the following:

$$\frac{\partial C}{\partial t} = \left[J(x,t) - J(x,t) - dx\frac{\partial J(x,t)}{\partial x} \right](dx)^{-1} = -\frac{\partial J(x,t)}{\partial x}, \tag{12.26}$$

and by making use of Fick's first law we obtain

$$\frac{\partial C}{\partial t} = \frac{\partial D(\partial C/\partial x)}{\partial x} \tag{12.27}$$

For the case where the diffusion coefficient is independent of x we have

$$\frac{\partial C}{\partial t} = D\frac{\partial^2 C}{\partial x^2}. \tag{12.28}$$

At this point we have obtained two very basic laws for describing transport of particles in solution. Later we will see applications of these laws for describing a chemical reaction in solution.

12.2 BROWNIAN MOTION

Here we consider a particle of mass m with a center of mass coordinate $x(t)$ at time t and a corresponding velocity v. The particle is immersed in a solution having a temperature T. The solution has, of course, an extremely large number of degrees of freedom and we will not describe in detail here the interaction of the center of mass coordinate x with the numerous degrees of freedom of the solution. Alternatively, the solvent's degrees of freedom can be regarded as constituting a heat reservoir at some temperature T and interaction of the solvent molecules with our particle or solute with a center of mass coordinate can be described by some effective force $F(t)$ (see Fig. 12.4).

If the solute interacts with some external field, gravitational or electromagnetic, there is an additional force $F_{ex}(t)$. We are then able to write Newton's second law of motion for the center of mass coordinate as

$$m \frac{dv}{dt} = F_{ex}(t) + F(t),$$ (12.29)

where $F(t)$ depends on the position of the solvent molecules that are in constant motion. The force, $F(t)$, is generally a rapidly fluctuating function of time and its variation is highly irregular. The force $\langle F(t) \rangle$ is a mean or average force sampled over all the different ensembles at a given time t (see Appendix D on Statistical Mechanics). The average force, $\langle F(t) \rangle$, is obtained as

$$\langle F(t) \rangle = (1/N) \sum_{k=1}^{N} F_k(t)$$ (12.30)

That which is stated in Eq. (12.9) is assumed to hold for each member of the ensemble, that is for each k-ensemble. N is the total number of ensembles. The variation of $F(t)$ is characterized by a correlation time, τ, which is defined as the mean time between two successive maxima of the fluctuating force $F(t)$. As $F(t)$ is a rapidly fluctuating force, the velocity v of our particle will fluctuate rapidly. We will assume that the velocity can be written as

$$v(t) = \langle v(t) \rangle + v(t)^*$$ (12.31)

where $\langle v \rangle$ is an average velocity for the solute and a slowly varying function of time, whereas v^* is the part of the velocity that fluctuates rapidly. The average velocity is of crucial importance in determining the behavior of the particle over long periods of time. The external force, $F_{ex}(t)$, varies very much more slowly than $F(t)$ and can be taken to be a constant force during the time interval t.

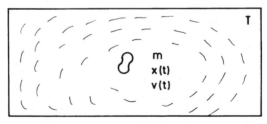

Fig. 12.4. A Brownian particle in a solution at temperature T [mass m, position $x(t)$, velocity $v(t)$].

The interaction force, $F(t)$, must possess the property that, in the absence of an external force and with an average velocity different from zero, the system reaches an equilibrium state for which $\langle v \rangle$ is zero. The force, $F(t)$, is affected by the motion of the particles in such a way that $F(t)$ is made up of a slowly varying part, $\langle F(t) \rangle$, that tends to restore the particle to equilibrium and a rapidly fluctuating part, $F(t)^*$,

$$F(t) = \langle F(t) \rangle + F(t)^* \tag{12.32}$$

The slowly varying part, $\langle F(t) \rangle$, must be some function of $\langle v(t) \rangle$ such that in equilibrium, when $\langle v(t) \rangle = 0$, $\langle F(t) \rangle = 0$. This can be recast as

$$\langle F(t) \rangle = -\alpha \langle v(t) \rangle \tag{12.33}$$

where α is a positive constant, referred to as the friction constant. The minus sign in Eq. (12.33) indicates explicitly that the force acts in a direction which tends to reduce the velocity to zero as time increases. Our equation of motion is given by

$$m\frac{dv}{dt} = F_{ex}(t) - \alpha[\langle v(t) \rangle + v(t)^*] + F(t)^*$$
$$= F_{ex}(t) - \alpha v(t) + F(t)^* \tag{12.34}$$

which is the Langevin equation, where we have expressed the force in terms of (i) a slowly varying part $-\alpha v(t)$ and (ii) a rapidly fluctuating part $F(t)^*$, purely random and having a mean value that vanishes. The Langevin equation describes in this way the behavior of the particle at all later times, if its initial conditions are specified.

As an illustration we will calculate the mean-square displacement in the absence of an external field

$$m\frac{dv}{dt} = -\alpha v(t) + F(t)^* \tag{12.35}$$

At thermal equilibrium $\langle x(t) \rangle = 0$, since by symmetry there is no preferred direction for the particle to move. We calculate $\langle x(t)^2 \rangle$ of the particle in a time interval t

$$mx\frac{d^2x}{dt^2} = m\left[\frac{d}{dt}\left(x\frac{dx}{dt}\right) - \left(\frac{dx}{dt}\right)^2\right] = -\alpha x\frac{dx}{dt} + xF(t)^* \tag{12.36}$$

By averaging Eq. (12.36) we obtain

$$m\left[\left\langle \frac{d}{dt}\left(x\frac{dx}{dt}\right)\right\rangle - \left\langle \left(\frac{dx}{dt}\right)^2\right\rangle\right]$$
$$= -\alpha\left\langle x\frac{dx}{dt}\right\rangle + \langle xF(t)^*\rangle \tag{12.37}$$

From the definition of $F(t)^*$ we know that $\langle F(t)^*\rangle = 0$ which holds independently of x or v. Making use of the equipartition theorem (see Chapter 5) we can write

$$\frac{1}{2}m\left\langle \left(\frac{dx}{dt}\right)^2\right\rangle = \frac{1}{2}kT \tag{12.38}$$

and this brings Eq. (12.37) into the following form:

$$m\left\langle \frac{d}{dt}\left(x\frac{dx}{dt}\right)\right\rangle = kT - \alpha\left\langle x\frac{dx}{dt}\right\rangle. \tag{12.39}$$

Formally we can write the solution as

$$\left\langle x\frac{dx}{dt}\right\rangle = \frac{kT}{\alpha} + Ce^{-\gamma t} \tag{12.40}$$

where C is a constant specified by the initial conditions and γ is equal to α/m. Next, by making use of the initial condition, which states that at $t = 0$, $x = 0$, we obtain $C = -kT/\alpha$. This in turn gives

$$\left\langle x\frac{dx}{dt}\right\rangle = \frac{1}{2}\frac{d\langle x^2\rangle}{dt} = \frac{kT}{\alpha}(1 - e^{-\gamma t}) \tag{12.41}$$

Integration of Eq. (12.41) gives

$$\langle x^2\rangle = \frac{2kT}{\alpha}\left[t - \frac{1}{\gamma}(1 - e^{-\gamma t})\right] \tag{12.42}$$

Let us consider two limiting cases: (i) $t \ll 1/\gamma$ and (ii) $t \gg 1/\gamma$. In the first case we obtain

$$\langle x^2 \rangle \approx \frac{2kT}{\alpha} \left\{ t - \frac{1}{\gamma} \left[1 - \left(1 - \gamma t + \frac{1}{2} (\gamma t)^2 \right) \right] \right\} = \frac{kT}{m} t^2, \qquad (12.43)$$

which is the equation describing a free particle with constant thermal velocity $v = \sqrt{kT/m}$. In the second case we obtain

$$\langle x^2 \rangle = \frac{2kTt}{\alpha} \qquad (12.44)$$

In this case the particle behaves like a diffusing particle performing a random walk, where $\langle x^2 \rangle$ is proportional to t. The diffusion equation (12.15) allows us to write

$$\langle x^2 \rangle = 2Dt, \qquad (12.45)$$

which means that our diffusion coefficient is given by

$$D = \frac{kT}{\alpha} \qquad (12.46)$$

In the case of a spherical particle the friction coefficient is given as (see ref. [2])

$$\alpha = 6\pi\eta a \qquad (12.47)$$

where a is the radius of the particle and η is the viscosity of the solvent. The displacement thus becomes

$$\langle x^2 \rangle = \frac{kTt}{(3\pi\eta a)} \qquad (12.48)$$

Lastly we will consider a charged particle (q) in an external field E for which we obtain

$$m \left\langle \frac{dv}{dt} \right\rangle = qE - \alpha\langle v(t) \rangle + \langle F(t)^* \rangle \qquad (12.49)$$

and for the steady state, that is for $\langle dv/dt \rangle = 0$ we have

$$0 = qE - \alpha\langle v(t) \rangle \qquad (12.50)$$

We define the mobility, μ, as

$$\mu = \frac{\langle v(t) \rangle}{E} \tag{12.51}$$

We then obtain, using the definition of D, the Einstein relation

$$\frac{\mu}{D} = \frac{q}{kT} \tag{12.52}$$

12.3 AN APPLICATION OF FICK'S LAW

As an illustration let us consider a one-dimensional reaction coordinate along the x-axis. This involves a compound p that moves along the positive part of the x-axis. An electrode is positioned at $x = 0$, reducing the compound p and thereby removing any p that reaches the electrode, thereby assuming that the electron transfer between the electrode and the compound p is infinitely fast. The electrode is turned on at $t = 0$. The concentration of p is given by $C_p(x, t)$ and is determined through Fick's law, which states

$$\frac{\partial C_p(x, t)}{\partial t} = D \frac{\partial^2 C_p(x, t)}{\partial x^2} \tag{12.53}$$

with the initial condition that $C_p(x, 0)$ is equal to the bulk concentration C^*, and that for x far from the surface of the electrode, the concentration $C_p(x, t)$ is also equal to C^*:

$$C_p(x, 0) = C^* \quad \text{and} \quad \lim_{x \to \infty} C_p(x, t) = C^* \tag{12.54}$$

In order to solve this problem we introduce the Laplace transform in t of the function $F(t)$ (see Appendix C on Laplace transformation)

$$L\{F(t)\} = \int_0^\infty e^{-st} F(t)\, dt = f(s) \tag{12.55}$$

and from that

$$L\left\{\frac{dF(t)}{dt}\right\} = \int_0^\infty e^{-st} \frac{d}{dt} F(t)\, dt$$

$$= [e^{-st}F(t)]_0^\infty + s \int_0^\infty e^{-st}F(t)\, dt$$

$$= -F(0) + s \int_0^\infty e^{-st}F(t)\, dt = -F(0) + s f(s) \qquad (12.56)$$

and also

$$L\left\{\frac{\partial F(x, t)}{\partial x}\right\} = \frac{\partial}{\partial x} f(x, s) \qquad (12.57)$$

The Laplace transform of the right hand side of Eq. (12.53) gives

$$L\left\{\frac{\partial C_p(x, t)}{\partial t}\right\} = s c_p(x, s) - C^* \qquad (12.58)$$

where $c_p(x, s)$ is the Laplace transform of $C_p(x, t)$. The Laplace transform of the left-hand side of Eq. (12.53) gives

$$L\left\{D\frac{\partial^2 C_p(x, t)}{\partial x^2}\right\} = D\frac{d^2 c_p(x, s)}{dx^2} \qquad (12.59)$$

where we have rewritten Eq. (12.53) as

$$s c_p(x, s) - C^* = D\frac{d^2 c_p(x, s)}{dx^2}. \qquad (12.60)$$

A solution to Eq. (12.60) is given by

$$c_p(x, s) = \frac{C^*}{s} + A(s)\exp[-x(s/D)^{1/2}] + B(s)\exp[-x(s/D)^{1/2}] \qquad (12.61)$$

for which we can determine the two coefficients, $A(s)$ and $B(s)$, using the boundary conditions at $x = 0$ and $x = \infty$. Eq. (12.61) is thus reduced to

$$c_p(x, s) = \frac{C^*}{s}(1 - \exp[-x(s/D)^{1/2}]) \qquad (12.62)$$

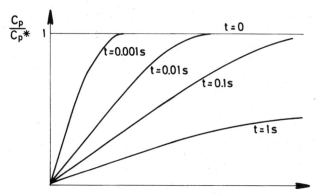

Fig. 12.5. Concentration profiles as a function of time and distance from the surface of the electrode.

Finally we obtain, by performing the inverse Laplace transformation (see Table C.1 in Appendix C),

$$C_p(x,t) = L^{-1}\{c_p(x,s)\} = C^*[1 - \mathrm{erfc}(x/2\sqrt{Dt})] \qquad (12.63)$$

where $\mathrm{erfc}(z)$ is the error function.

Fig. 12.5 shows the normalized concentration profiles as a function of time and distance from the surface of the electrode. Note the sharp gradient in the concentration with respect to x for small t. For larger t it levels off towards a gradient that is given by the limitation of mass transfer.

REFERENCES

[1] F. Reif, "Fundamentals of Statistical and Thermal Physics," McGraw-Hill, New York, 1965.

[2] J. O. Hirschfelder, C. F. Curtiss and R. B. Bird, "Molecular Theory of Gases and Liquids," John Wiley & Sons, New York, 1954.

EXERCISES

1. Derive Eq. (12.15).

2. Derive the Gaussian distribution probability function (12.17).

3. The mobility, u_j, is related to the diffusion coefficient, D_j by the equation $u_j = z_j FD_j/RT$, where z_j is the numerical charge, F Faraday's constant, R the gas constant, and T the temperature. From the mobility data, calculate

the diffusion coefficient and explain the differences between the three ions H^+, I^-, and Li^+.

$$u(H^+) = 3.62510^{-3} \text{ cm}^2 \text{ sec}^{-1} \text{ V}^{-1}$$
$$u(Li^+) = 4.01010^{-4} \text{ cm}^2 \text{ sec}^{-1} \text{ V}^{-1}$$
$$u(I^-) = 7.9610^{-4} \text{ cm}^2 \text{ sec}^{-1} \text{ V}^{-1} \quad (12.64)$$

The temperature is $25°$ C.

4. Consider a liquid containing a microscopic particle A, having a mass m and a velocity v. The temperature of the system is T. The equation of motion (EOM) is

$$m \frac{dv}{dt} = -\gamma v + F(t) \quad (12.65)$$

where γ and $F(t)$ are the friction constant and the fluctuating part of the force, respectively. The fluctuating force is a stochastic variable with the properties

$$\langle F(t) \rangle = 0$$
$$\langle F(t)F(t') \rangle = 2\gamma k T \delta(t - t') \quad (12.67)$$

Solve the EOM formally for t in the interval $[t_0; t]$ and discuss the solution. Calculate the ensemble average of $v(t)$ (make use of $\langle F(t) \rangle$) and discuss the relaxation process towards equilibrium.
Finally show that

$$\langle v(t)v(t') \rangle = \frac{1}{m^2} \int_{-\infty}^{t} ds \exp[-(t-s)/\tau] \int_{-\infty}^{t'} ds'$$
$$\cdot \exp[-(t'-s')/\tau]2\gamma k T \delta(s-s')$$

$$(12.68)$$

Further algebraic manipulation will give

$$\langle v(t)v(t') \rangle = A f(-|t-t'|/\tau). \quad (12.69)$$

Determine A and the function f.

13

ENERGETIC ASPECTS OF SOLVENT EFFECTS ON SOLUTES

13.1 INTRODUCTION

A molecule immersed in a solvent is affected by the solvent, and the molecular properties of the molecule (the solute) are expected to be different from those of the molecule in the vacuum. Interactions between the solvent and the solute can be both long- and short-range. The first theorectical studies of these effects considered solvation energies [1–4]. Subsequent developments have spurred further investigations, making the calculation of solvent-induced effects of molecular properties an increasingly active research area. The different models encountered in the literature can be divided into three types; (i) supermolecular models; (ii) continuum models; and (iii) semicontinuum models. In the first type one considers a molecule solvated by a limited number of solvent molecules that form a cluster around the solute. This approach enables a detailed description of the solvent effects due to the first solvation shell, but long-range interactions are poorly described. The basic idea of the type (ii) models is that the solvated molecule induces polarization charges in an outer medium and that these give rise to an extra potential at the position of the solvated molecule. The response field or reaction field from the induced polarization charges thus interacts with the molecular system and gives rise to a solvent effect. This model enables a proper electrostatic description of the long-range Coulomb interactions. Type (iii) models are a hybrid of the two other models; they involve a cluster of solvent molecules around the solute and these solvent molecules are themselves surrounded by a dielectric medium.

In the present chapter we will consider models of types (ii) and (iii) and illustrate the assumptions underlying these models.

13.2 THE DIELECTRIC MEDIUM REPRESENTATION

Let us consider a solute, represented by its molecular charge distribution, enclosed in a spherical cavity that is, in turn, embedded in a structureless polarizable dielectric medium described by macroscopic bulk dielectric constants. The molecular charge distribution inside the cavity induces polarization charges in the dielectric medium. This in turn leads to a reaction field and a potential in the dielectric (V_{pol}) and the induced field interacts with the charge distribution $\rho(\mathbf{r})$ at a point $\mathbf{r}$ inside the cavity (see Fig. 13.1). The resulting polarization energy is given by

$$E_{pol} = \tfrac{1}{2} \int d\mathbf{r}\, \rho(\mathbf{r}) V_{pol}(\mathbf{r}) \tag{13.1}$$

where

$$4\pi\epsilon_o V_{pol}(\mathbf{r}) = \int da' \frac{\sigma_p(\mathbf{r}')}{|\mathbf{r} - \mathbf{r}'|} + \int d\mathbf{r}' \frac{\rho_p(\mathbf{r}')}{|\mathbf{r} - \mathbf{r}'|} \tag{13.2}$$

and σ_p and ρ_p are the surface and volume polarization charge densities

$$\sigma_p = \mathbf{P} \cdot \mathbf{n} \tag{13.3}$$

$$\rho_p = div\,\mathbf{P} = \frac{\partial P_x}{\partial x} + \frac{\partial P_y}{\partial y} + \frac{\partial P_z}{\partial z} \tag{13.4}$$

$$\mathbf{n} = -\frac{\mathbf{r}'}{|\mathbf{r}'|} \tag{13.5}$$

Here $\mathbf{P} = (P_x, P_y, P_z)$ is the polarization of the medium and $\mathbf{n}$ the outwardly directed normal to the cavity surface. The charge distribution ρ is assumed to

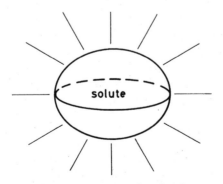

Fig. 13.1. A solvated molecule.

be zero outside the cavity. The surface and volume integrations are denoted da' and dr', respectively.

The dielectric polarization of a linear isotropic medium is related to the electric field vector $\mathbf{E}$ through the following relation

$$\mathbf{P}(\mathbf{r}) = \chi \mathbf{E}(\mathbf{r}) \tag{13.6}$$

The electric susceptibility χ is related to the dielectric constant (κ) in the following way:

$$\kappa = 1 + \frac{\chi}{\epsilon_o} \tag{13.7}$$

where ϵ_0 is the vacuum permittivity. The total electrostatic potential is the sum of the potential from the charge distribution in the cavity and the polarization potential from the induced charges in the continuum given by Eq. (13.2):

$$V(\mathbf{r}) = V_{\text{pol}}(\mathbf{r}) + \frac{1}{4\pi\epsilon_o} \int dr' \frac{\rho(\mathbf{r}')}{|\mathbf{r} - \mathbf{r}'|} \tag{13.8}$$

The following integral equation may now be established for the polarization field, using Eq. (13.6) and the relation $\mathbf{E} = -\nabla V$:

$$4\pi\epsilon_o \mathbf{P}(\mathbf{r}) + \chi\nabla \int da' \frac{\mathbf{P}(\mathbf{r}') \cdot \mathbf{n}'}{|\mathbf{r} - \mathbf{r}'|} = -\chi\nabla \int dr' \frac{\rho(\mathbf{r}')}{|\mathbf{r} - \mathbf{r}'|} \tag{13.9}$$

We have also used the fact that the medium is isotropic (i.e., b is the radius of the cavity):

$$div\,\mathbf{P}(\mathbf{r}) = 0, \qquad |\mathbf{r}| > b\,(|\mathbf{r}| > |\mathbf{r}'|) \tag{13.10}$$

We then get the polarization field from the integral equation, and the polarization energy is determined (see Appendix E) to be

$$E_{\text{pol}} = -\frac{1}{2} \sum_{\lambda\mu} \Phi^\dagger_{\lambda\mu} M_{\lambda\mu} = -\frac{1}{2} \sum_{\lambda\mu} (-1)^\mu f_\lambda M_{\lambda-\mu} M_{\lambda\mu} \tag{13.11}$$

The following relations apply:

$$M_{\lambda\mu} = \int d\mathbf{r}\,\rho(\mathbf{r}) S_{\lambda\mu}(\mathbf{r}), \tag{13.12}$$

$$S_{\lambda\mu}(\mathbf{r}) = \sqrt{\frac{4\pi}{(2\lambda + 1)}} \, r^{\lambda} Y_{\lambda\mu}(\theta, \phi), \tag{13.13}$$

$$\Phi_{\lambda\mu} = f_{\lambda} M_{\lambda\mu}, \tag{13.14}$$

and

$$4\pi\epsilon_o f_{\lambda} = b^{-(2\lambda+1)} \frac{(\lambda + 1)(\kappa - 1)}{\lambda + \kappa(\lambda + 1)} \tag{13.15}$$

where $Y_{\lambda\mu}(\theta, \phi)$ are the spherical harmonics.

Here $M_{\lambda\mu}$ are the charge multipole densities, $S_{\lambda\mu}$ the spherical polynomials, and $\Phi_{\lambda\mu}$ the $\lambda\mu$-component of the reaction field. In what follows, we will switch to atomic units, which are more convenient for the purpose of calculation.

The molecular charge distribution inside the cavity is calculated in the following manner:

$$\rho(\mathbf{r}) = \langle \Psi | \rho_{\mathrm{mol}}(\mathbf{r}) | \Psi \rangle \tag{13.16}$$

where Ψ is the electronic wave function of the solute and $\rho_{\mathrm{mol}}(\mathbf{r})$ the density operator,

$$\rho_{\mathrm{mol}}(\mathbf{r}) = \sum_k Z_k \, \delta(\mathbf{r} - \mathbf{R}_k) - \sum_j \delta(\mathbf{r} - \mathbf{r}_j) \tag{13.17}$$

Here Z_k is the charge of nucleus k, and $\mathbf{R}_k$ and $\mathbf{r}_j$ are the coordinates of nucleus k and electron j, respectively. The summations are over the full set of nuclei and electrons.

The dielectric solvation energy for the molecular charge distribution embedded in a linear isotropic dielectric medium is the product of the expectation values for the reaction field and the multipole charge moments

$$E_{\mathrm{pol}} = -\frac{1}{2} \sum_{\lambda\mu} \langle \Phi_{\lambda\mu}^{\dagger} \rangle \langle M_{\lambda\mu} \rangle \tag{13.18}$$

$$\langle M_{\lambda\mu} \rangle = \sum_k Z_k S_{\lambda\mu}(\mathbf{R}_k) - \langle S_{\lambda\mu} \rangle \tag{13.19}$$

$$\langle \Phi_{\lambda\mu} \rangle = f_{\lambda} \langle M_{\lambda\mu} \rangle \tag{13.20}$$

$$f_{\lambda} = b^{-(2\lambda+1)} \frac{(\lambda + 1)(\kappa - 1)}{\lambda + \kappa(\lambda + 1)} \tag{13.21}$$

The expression for E_{pol} in Eq. (13.18) contains the electrostatic and polarization interactions between the solute and the dielectric medium, but not the dispersion energy. Until now we have only considered one type of polarization vector, which is sufficient for investigating equilibrium problems (where the solvent response is in equilibrium with the charge distribution of the solute).

13.3 NONEQUILIBRIUM SOLVENT CONFIGURATION

Investigations of nonequilibrium problems, where the solvent response is not in equilibrium with the charge distribution of the solute, require at least two different polarization vectors. A generalization of the McRae-Bayliss model [5,6] considers the polarization vector, $\mathbf{P}$, to be a sum of the optical polarization vector, $\mathbf{P}_{op}$, and the inertial polarization vector, $\mathbf{P}_{in}$,

$$\mathbf{P} = \mathbf{P}_{op} + \mathbf{P}_{in} \qquad (13.22)$$

The optical polarization vector is always in equilibrium with the molecular charge distribution due to its very short relaxation time; this polarization is assumed to change instantaneously with the changes in the molecular charge distribution. The optical polarization represents the response from the electronic degrees of freedom of the outer solvent.

The inertial polarization represents the response due to the nuclear degrees of freedom of the solvent molecules and the solvent molecular motion, i.e. both rotational and translational. This type of polarization should, in a more elaborate description of the outer solvent, be divided into different types of polarization with significantly different relaxation times. For typical solvents, assuming a Debye-type response (single exponential decay), the inertial polarization has relaxation times on the order of nano- or picoseconds. This is a much longer time scale than that for electronic excitations. Therefore it is physically reasonable to assume that during an electronic excitation, the inertial polarization remains fixed and corresponds to the molecular charge distribution of the initial state. This assumption is completely analogous to the so-called Franck-Condon principle and sudden approximations involving frozen innershell electrons frequently used in analysis of electronic spectra for compounds in vacuum.

The physical picture of the model is one where the solute and optional solvation shells are enclosed in a spherical cavity immersed in a dielectric medium with two types of polarization, optical and inertial. The electronic excitation of the molecular complex within the cavity is represented by two electronic wave functions, Ψ_i and Ψ_f, that are the electronic wave functions for the initial and final state, respectively. The initial state corresponds to the situation where the molecular charge distribution and the inertial polarization are in equilibrium with each other. Therefore

$$\mathbf{P}_{op}^i = \mathbf{P}_{op}(\rho_i) \qquad (13.23)$$

and

$$\mathbf{P}_{in}^i = \mathbf{P}_{in}(\rho_i), \tag{13.24}$$

where ρ_i is the charge distribution of the molecular complex in the initial state corresponding to the electronic wave function Ψ_i, and $\mathbf{P}_{op}^i$ and $\mathbf{P}_{in}^i$ are the optical and inertial polarization vectors for the initial state.

For the final state, having the electronic wave function Ψ_f, it is only the optical polarization that is in equilibrium with the charge distribution of the final state. The inertial polarization of the final state has the same value as in the initial state, since it has not been able to change because the electronic excitation has been so fast. Therefore

$$\mathbf{P}_{op}^f = \mathbf{P}_{op}(\rho_f) \tag{13.25}$$

and

$$\mathbf{P}_{in}^f = \mathbf{P}_{in}(\rho_i), \tag{13.26}$$

where ρ_f is the charge distribution of the molecular complex in the final state corresponding to the electronic wave function Ψ_f. The two polarization vectors, $\mathbf{P}_{op}^f$ and $\mathbf{P}_{in}^f$, correspond to the optical and inertial polarizations for the final state. The dielectric constant for the optical polarization vector is ϵ_{op} and the dielectric constant for the total polarization vector is ϵ_{st}.

The total energy of a given state is

$$E = E_{vac} + E_{pol} \tag{13.27}$$

where E_{vac} is the energy of the system in vacuum.

The total energy of the initial state, E_1, is given as

$$E_1 = E_{1,vac} - \frac{1}{2} \sum_{\lambda\mu} \langle \Phi_{\lambda\mu}^\dagger \rangle \langle M_{\lambda\mu}(\rho_i) \rangle \tag{13.28}$$

where $\langle M_{\lambda\mu}(\rho_i) \rangle$ is the charge distribution of the solute in the initial state and $\langle \Phi_{\lambda\mu}^\dagger \rangle$ is the polarization field from the solvent. The latter is given as

$$\begin{aligned}
\langle \Phi_{\lambda\mu} \rangle &= f_\lambda(\epsilon_{op}) \langle M_{\lambda\mu}(\rho_i) \rangle \\
&\quad + f_\lambda(\epsilon_{st}, \epsilon_{op}) \langle M_{\lambda\mu}(\rho_i) \rangle \\
&= f_\lambda(\epsilon_{st}) \langle M_{\lambda\mu}(\rho_i) \rangle
\end{aligned} \tag{13.29}$$

where ϵ_{op} is the optical and ϵ_{st} the static dielectric constants and where

$$f_\lambda(\epsilon) = b^{-(2\lambda+1)} \frac{(\lambda+1)(\epsilon-1)}{\lambda + \epsilon(\lambda+1)} \tag{13.30}$$

and

$$f_\lambda(\epsilon_{st}, \epsilon_{op}) = b^{-(2\lambda+1)} \frac{(\lambda+1)(\epsilon_{st}-1)}{\lambda + \epsilon_{st}(\lambda+1)}$$
$$- b^{-(2\lambda+1)} \frac{(\lambda+1)(\epsilon_{op}-1)}{\lambda + \epsilon_{op}(\lambda+1)} \qquad (13.31)$$

For the final state we obtain the following energy

$$E_2 = E_{2,\text{vac}} - \frac{1}{2} \sum_{\lambda\mu} \langle \Phi_{\lambda\mu}^\dagger \rangle \langle M_{\lambda\mu}(\rho_f) \rangle \qquad (13.32)$$

where $\langle M_{\lambda\mu}(\rho_f) \rangle$ is the charge distribution of the solute in the final state and $\langle \Phi_{\lambda\mu}^\dagger \rangle$ is the polarization field from the solvent. According to Eqs. (13.23–13.26) the solvent response is given as

$$\langle \Phi_{\lambda\mu} \rangle = f_\lambda(\epsilon_{op}) \langle M_{\lambda\mu}(\rho_f) \rangle + f_\lambda(\epsilon_{st}, \epsilon_{op}) \langle M_{\lambda\mu}(\rho_i) \rangle \qquad (13.33)$$

where the expressions for $f_\lambda(\epsilon_{op})$ and $f_\lambda(\epsilon_{st}, \epsilon_{op})$ are as given in Eqs. (13.30) and (13.31).

The energy difference between the initial state and the final state is then

$$\Delta E = E_2 - E_1 = E_{2,\text{vac}} - E_{1,\text{vac}}$$
$$- \frac{1}{2} \sum_{\lambda\mu} (f_\lambda(\epsilon_{op}) \langle M_{\lambda\mu}(\rho_f)^\dagger \rangle \langle M_{\lambda\mu}(\rho_f) \rangle$$
$$+ f_\lambda(\epsilon_{st}, \epsilon_{op}) \langle M_{\lambda\mu}(\rho_i)^\dagger \rangle \langle M_{\lambda\mu}(\rho_f) \rangle)$$
$$+ \frac{1}{2} \sum_{\lambda\mu} f_\lambda(\epsilon_{st}) \langle M_{\lambda\mu}(\rho_i)^\dagger \rangle \langle M_{\lambda\mu}(\rho_i) \rangle \qquad (13.34)$$

where $\dagger$ denotes complex conjugation. We can rewrite this energy difference as

$$\Delta E = E_{2,\text{vac}} - E_{1,\text{vac}}$$
$$- \frac{1}{2} \sum_{\lambda\mu} f_\lambda(\epsilon_{op})(\langle M_{\lambda\mu}(\rho_f)^\dagger \rangle \langle M_{\lambda\mu}(\rho_f) \rangle - \langle M_{\lambda\mu}(\rho_i)^\dagger \rangle \langle M_{\lambda\mu}(\rho_i) \rangle)$$
$$- \frac{1}{2} \sum_{\lambda\mu} f_\lambda(\epsilon_{st}, \epsilon_{op}) \langle M_{\lambda\mu}(\rho_i)^\dagger \rangle (\langle M_{\lambda\mu}(\rho_f) \rangle - \langle M_{\lambda\mu}(\rho_i) \rangle) \qquad (13.35)$$

As an illustration let us consider a solute in two different solvents with static dielectric constants ϵ_{st}^a and ϵ_{st}^b. We assume that the optical dielectric constants for the two solvents are the same and that the charge distributions are given by dipole moments that do not change when changing the solvent. We obtain as the change in the energy difference with respect to changing the solvent

$$\delta_{ab}(\Delta E) = -\frac{1}{2} f_1(\epsilon_{st}^b, \epsilon_{st}^a) \mu(\mu^* - \mu) \qquad (13.36)$$

where

$$f_1(\epsilon_{st}^b, \epsilon_{st}^a) = b^{-3} \left[\frac{2(\epsilon_{st}^b - 1)}{1 + 2\epsilon_{st}^b} - \frac{2(\epsilon_{st}^a - 1)}{1 + 2\epsilon_{st}^a} \right] \qquad (13.37)$$

The two dipole moments μ and μ^* correspond to the initial and final states, respectively. These equations have been used for relating shifts in absorption energies of solutes in a series of different solvents [5,6].

Recent revelations in the area of nonequilibrium solvent states have greatly improved the understanding of how solutes are perturbed [7].

REFERENCES

[1] M. Born, Z. Phys. **1**,45(1920).

[2] J. G. Kirkwood, J. Chem. Phys. **2**,351(1934).

[3] J. G. Kirkwood and F. Westheimer, J. Chem. Phys. **6**,506(1936).

[4] L. Onsager, J. Am. Chem. Soc. **58**,1486(1936).

[5] E. McRae, J. Phys. Chem. **61**,562(1957).

[6] N. Bayliss and E. McRae, J. Phys. Chem. **58**,1002(1954).

[7] H. J. Kim and J. T. Hynes, J. Chem. Phys. **96**,5088(1992).

[8] Y. Ooshika, J. Phys. Soc. Jap. **9**,594(1954).

[9] E. Lippert, Z. Naturforsch **10a**,541(1955).

[10] N. Mataga, Y. Kaifu and M. Koizumi, Bull. Chem. Soc. Jap. **29**,465(1956).

EXERCISES

1. For a solute enclosed by a spherical cavity with radius a and a static dielectric constant of ϵ_{st}, determine the dielectric solvation energy and discuss the difference in dependencies of the molecular radius a when the solute is:

   a. An ion with charge q.

   b. A dipole with dipole moment μ_{10}.

   c. A quadrupole with moment Q_{20}.

2. Influence of the solvent on absorption and fluorescence spectra has in many occasions been approximated by the Ooshika-Lippert-Mataga equation [8–10]:

$$\langle \Delta \nu \rangle = \langle \Delta \nu \rangle_{abs} - \langle \Delta \nu \rangle_{flu} = \frac{2(\mu_{ex} - \mu_{gs})^2}{hca^3} \left[\frac{\epsilon_{st} - 1}{2\epsilon_{st} + 1} - \frac{\epsilon_{op} - 1}{2\epsilon_{op} + 1} \right] \qquad (13.38)$$

In this equation $\langle \Delta \nu \rangle_{abs}$ and $\langle \Delta \nu \rangle_{flu}$ are the average solvent-induced frequency shifts of the absorption and fluorescence transitions, respectively. The solute dipole moments of the ground state and the excited state are given by μ_{gs} and μ_{ex}, respectively. The static dielectric constant of the solvent is ϵ_{st} and the optical dielectric constant is ϵ_{op}. The speed of light is c and Planck's constant is h. Derive the Ooshika-Lippert-Mataga equation and discuss its limitations.

3. We consider electronic excitations of formaldehyde (H_2CO) in different solvents. We will use data from Table 13.1.

TABLE 13.1. Electronic Excitations of Formaldehyde in Different Solvents Relative to Ethyl Ether (atomic units)

Solvent	$GS \to A_2$	$GS \to B_1$	$GS \to B_2$
Benzene	−0.25841	−0.0208	−0.01014
Ethyl ether	0	0	0
1-Hexanol	0.03313	0.01963	0.00875
Acetone	0.10015	0.02301	0.01019
Methanol	0.13218	0.02531	0.01114
Water	0.13882	0.02795	0.01225

a. Consider whether it is valid to make use of Eqs. (13.36) and (13.37) for the three different electronic excitations: $GS \to A_2$, $GS \to B_1$, and $GS \to B_2$. GS denotes the ground state and the other terms denote the symmetry of the first excited state.

b. Determine the dipole moments for the excited states (only for the transitions where the two equations hold) in terms of the dipole moment for the ground state.

c. For a given solvent, determine the fully solvent-relaxed excited states from (b) in terms of solvent and ground state properties. As solvents or dielectric media we have chosen benzene ($\epsilon_{st} = 2.284$, $\epsilon_{op} = 2.244$), ethyl ether ($\epsilon_{st} = 4.335$, $\epsilon_{op} = 1.828$), 1-hexanol ($\epsilon_{st} = 13.3$, $\epsilon_{op} = 2.005$), acetone ($\epsilon_{st} = 20.7$, $\epsilon_{op} = 1.841$), methanol ($\epsilon_{st} = 32.63$, $\epsilon_{op} = 1.758$), and water ($\epsilon = 78.54$, $\epsilon_{op} = 1.778$). The spherical cavity has a radius of 2.645 Å.

14

MODELS FOR CHEMICAL
REACTIONS IN SOLUTION

In the next sections we will consider models for chemical reactions in solution (CRIS) and will cover some of the basic work by Eyring [1], Evans, Ogg, Polanyi [2,3], Smoluchowski [4], Debye [5], Onsager [6], Kramers [7], Marcus [8], and others. In the gas phase one could naively state that the reactants follow in isolation the route to products. On the other hand, in solution the solvent molecules continually perturb the reactants in their route to the products. In both cases we consider a reaction involving a multielectronic system with nuclear degrees of freedom, but in the case of CRIS we have, on top of that, a broad class of solvent degrees of freedom to take into account.

In recent years research on CRIS has become feasible by using somewhat detailed theoretical investigations. The development of analytical theory for liquid state structure and dynamics has improved the understanding of the supporting liquid medium, and computer experiments, molecular dynamics, and Monte Carlo simulations have verified and extended the understanding of the properties of liquid media. Naturally one would like to apply these ideas and techniques to chemical reactions in solution. Experimental investigations of CRIS have reached the stage of applying femto- and picosecond laser techniques to improve the understanding of the detailed mechanisms of CRIS.

Presently, there is no unified theory for CRIS: different reaction types and solvent classes require different approaches. The difficulties can be seen by considering schematically the time scales that are involved in CRIS (see Fig. 14.1). How do solvents influence chemical dynamics? At present we will state the obvious answers:

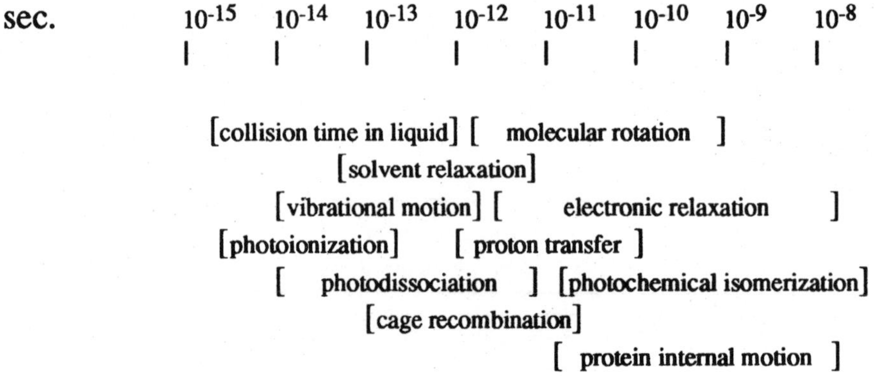

sec. 10^{-15} 10^{-14} 10^{-13} 10^{-12} 10^{-11} 10^{-10} 10^{-9} 10^{-8}
                 | | | | | | | |

[collision time in liquid] [molecular rotation   ]
[solvent relaxation]
[vibrational motion] [electronic relaxation   ]
[photoionization]   [proton transfer]
     [photodissociation   ] [photochemical isomerization]
          [cage recombination]
                    [protein internal motion]

Fig. 14.1. Time scales involved in chemical reactions in solution.

(1) They provide sources and sinks for energy. When reactants and products are separated by an energy barrier, solvents provide the energy necessary to reach the top of the barrier and enable the new product species to dispose of their excess energy.

(2) Solvents also create, enhance, induce, and impede molecular motion.

(3) They determine the energy surface on which the reactions take place.

(4) They provide a dielectric medium that can stabilize charged species. This affects the rate at which charge is transferred from one place to another.

One simple example of a solvent's impeding motion is the cage effect, first proposed in 1934 by J. Franck and E. Rabinowitch [9]. In a dissociation process

$$AB \rightarrow A + B \qquad (14.1)$$

there is a significant possibility that the surrounding solvent molecules will hinder the separation of the fragments A and B, in which case they then recombine. As we saw in Section 12.2 on Brownian motion, the concept of friction is merely a shortcut for describing the influence of the motion of neighboring molecules on the way a molecule moves and exchanges energy and momentum with its surroundings. In addition to collisional or viscous friction, which is responsible for impeding molecular motion, there is also dielectric friction. Dielectric friction is the response of the solvent to changes in the charge or charge distribution of the solute molecule; a lag in the solvent's ability to equilibrate a new charge can impose a frictional drag on an electron transfer reaction. Studies of rotational reorientation of molecules in solution have been particularly important in establishing theories and physical pictures of collisional friction. In general the rate of reorientation is inversely proportional to the viscosity (that is when the diameter of the solute molecule is larger than that of the solvent).

Chemical transformations are considerably slower than the molecular motion we have just discussed. At the level of an individual molecule, most reactions are rare events. At any one time only a small population of molecules are activated and participating in chemical reaction events. If the energy of the activated molecules must exceed that of the reactants the reaction rate will be proportional to $\exp(-E_{act}/kT)$. The fact that many activation energies, E_{act}, are much higher than the thermal energy kT explains the dramatic range of reaction rates. The scarcity of activated molecules is also a source of frustration in experimental kinetics because the crucial activated molecules cannot readily be studied due to their low concentration. By identifying activated molecules as those at a saddle point on the potential energy surface one could use the methods of equilibrium statistical mechanics to calculate their concentration. Molecules at a saddle point crossing toward the products are likely to continue their descent down the slope and form products. The transition state theory's Ansatz (or assumption) is that reactive trajectories cross through the saddle point once, and only once, on their way to the formation of products. This theory then gives an expression for rate constants involving only thermodynamical information about the transition state

$$k = \frac{w}{2\pi} e^{-F/kT} \qquad (14.2)$$

where w is the frequency of motion of reactants in the reactant potential well and F is the free energy of the barrier. Whether transition state theory (TST) is correct or not depends on the details of molecular motion.

Even for simple chemical reactions the condensed phase environment can change the way molecules move through the transition state. The friction and random forces arising from a condensed phase environment can affect the rare, activated trajectories just as they affect the ordinary, nonactivated molecular motion. There are two ways in which friction from the solvent affects motion through the transition state; these correspond to the high-friction limit and the low-friction limit. When friction is large the system will undergo Brownian or diffusive motion through the transition state and lead to many recrossings of the saddle point (see Fig. 14.2). Because TST counts each of these recrossings as a reaction event one must apply a correction factor, a transmission coefficient. This transmission coefficient, Π, is inversely proportional to the number of recrossings N_{rec}:

$$\Pi \approx \frac{1}{N_{rec}} \qquad (14.3)$$

As the friction becomes larger the number of recrossings gets larger, and Π gets smaller.

In the limit of little friction one has different types of behavior. Some friction is required to activate the system and to trap it after it has left the transition

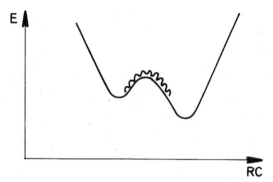

Fig. 14.2. Solvent effects on the transition state in the high-friction limit. RC denotes the reaction coordinate.

state (see Fig. 14.3). If no friction is present recrossings will occur. For very little friction the number of recrossings is inversely proportional to the friction constant, and in this regime the rate coefficient will increase as the friction increases.

An important concept has been introduced by R. Grote and J. T. Hynes [10], namely the frequency dependence of the friction. Their point is that motion through the transition state may well be so rapid that not all solvent fluctuation modes will be effective in damping the motion. In the moderate-to-large friction regime one can generalize H. A. Kramers' expression [7] in an elegant way: the expression remains intact but the friction coefficient in it is evaluated at a finite frequency. This frequency is the one associated with the unstable mode at the saddle point and must be self-consistently determined because it depends on the friction itself.

As indicated in Chapter 13, a condensed-phase environment can profoundly change the energetics of a chemical reaction. This is especially true when the

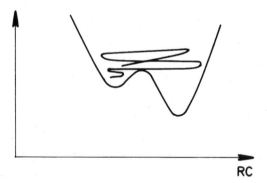

Fig. 14.3. Solvent effects on the transition state in the low-friction limit. RC denotes the reaction coordinate.

reaction involves the transfer of charge: for example, oxidation-reduction reactions or acid-base neutralization. The interaction energy of a charged species with a solvent can approach chemical bonding energies. In addition, the dynamics of solvation plays a role in charge transfer reactions. Because the interaction of a charge with the solvent environment is long range, solvation is a collective phenomenon. Since the phenomenon is collective and involves long-range forces, it is natural to use continuum theories to explain CRIS. The application of continuum theory to the solvation of ions was introduced by M. Born roughly 80 years ago [11]. The dynamical aspects of a dielectric medium during solvation or changes in charge distribution of a solute molecule were introduced by P. Debye [6] in terms of a frequency-dependent dielectric constant

$$\epsilon(w) = \epsilon_{op} + \frac{\epsilon_{op}\epsilon_{st}}{1 + iw\tau_D} \tag{14.4}$$

where i is the imaginary unit and where ϵ_{st} and ϵ_{op} are the static and optical dielectric constants for the medium, respectively. The time scale τ_D is on the order of the molecular rotation times. When solving the time-dependent continuum electrostatic problem, one discovers that the solvation polarization forms on a time scale referred to as the longitudinal relaxation time, $\tau_L = (\epsilon_{op}/\epsilon_{st})\tau_D$.

14.1 TRANSITION STATE THEORY

Basic assumptions of TST are the absence of (i) recrossing (ii) tunneling or quantum mechanical effects. The first assumption states that every trajectory crossing the transition state surface or barrier top from the side of reactants to the side of products must proceed towards products without recrossing. Naturally we will consider whether it is possible for the solvent to induce recrossings and as a simple example let us consider a collision complex A–B between the two atoms A and B, i.e.

$$A + B \rightleftharpoons A - B \tag{14.5}$$

In the gas phase we will have a rate constant according to TST which is

$$k_{gas}^{TST} = \frac{1}{\beta h Q_R} Q_{TS} e^{-E_{act}/kT} \tag{14.6}$$

and in the condensed phase we will have a similar expression except that the partition functions (Q_R for the reactants and Q_{TS} for the transition state) and the activation energy (E_{act}) are modified by the solvent. Note that $\beta = (kT)^{-1}$. The reaction in Eq. (14.5) has a reaction coordinate, which is the separation between the atoms A and B. The TST rate constant is given by the one-way flux $J_+(s)$

across the transition state surface averaged over an equilibrium distribution and normalized by the partition functions of the reactants:

$$k^{TST} = \langle J_+(r^*) \rangle_R \tag{14.7}$$

The transition state is located at r^* corresponding to a given A–B separation distance. The flux over the transition state "dividing line" at r^* is given by

$$J_+(r^*) = \delta(r - r^*)\frac{-p_r}{m}\,\theta(-p_r) \tag{14.8}$$

where the delta-function, $\delta(r - r^*)$, locates the flux at the transition state and the step function, $\theta(-p_r)$, restricts the relative momentum, p_r, to negative values so that the transition state is crossed. The reduced mass of the system is denoted by m and the total volume of the system is V. The "transition state" average of property A is given by

$$\langle A \rangle_R = \frac{V \int dr\,dp \int d\mathbf{r}^N\,d\mathbf{p}^N\,e^{(-\beta H^*)}A}{\int dr\,dp \int d\mathbf{r}^N\,d\mathbf{p}^N\,e^{(-\beta H_R)}\theta(r - r_c)} \tag{14.9}$$

where H^* is the Hamiltonian evaluated at the TST, H_R the reactant Hamiltonian, and r_c is a distance for which $U_{AB}(r > r_c) \sim 0$. The integration involves both the A–B relative coordinate and momentum (r, p), and the coordinates and momenta (r^N, p^N) of the N solvent molecules. The solvent molecules have a mass m_s and are otherwise not described in any detail. The step function $\theta(r - r_c)$ is 0 for $r < r_c$ and 1 for $r > r_c$. In this way we assure that the integration is done when there is no interaction between A and B, namely the situation involving the reactants. The separation distance indicates at what distance we are approaching the transition state area. Naturally we assume that this volume $(4\pi r_c^3/3V)$ is small and that we can neglect it. The Hamiltonian is schematically written as

$$H = H_{AB} + H_{sol} + V_{int} \tag{14.10}$$

where the first term on the right-hand side is the Hamiltonian for the reacting system A + B. In polar coordinates $\mathbf{r}$ is the separation vector, (θ, ϕ) are the polar angles for $\mathbf{r}$; and p_r, p_θ, and p_ϕ are the corresponding conjugate momenta. This term is then written

$$H_{AB} = \frac{1}{2m}\,[p_r^2 + r^{-2}(p_\theta^2 + \sin^{-2}\theta p_\phi^2)] + U_{AB}(r) \tag{14.11}$$

in which the first term is the kinetic energy and the second one the interaction energy between A and B. The second term in Eq. (14.10) is the Hamiltonian for the solvent and is expressed in terms of the kinetic and potential energy operators, $T(p^N)$ and $U(r^N)$, for the solvent. The last term in Eq. (14.10) represents the interaction operator between the solvent and the reacting complex. The rate constant is given as

$$k^{TST} = \langle J_+(\mathbf{r}^*) \rangle_R = \frac{V \int d\mathbf{r}\,d\mathbf{p} \int d\mathbf{r}^N\,d\mathbf{p}^N\,e^{(-\beta H^*)}\delta(r - r^*)\dfrac{-p_r}{m}\,\theta(-p_r)}{\int d\mathbf{r}\,d\mathbf{p} \int d\mathbf{r}^N d\mathbf{p}^N\,e^{(-\beta H)}\theta(r - r_c)} \qquad (14.12)$$

where we have dropped the subscript R for the reactant Hamiltonian (the step function secures that it is evaluated at a reactant configuration). Integrating over all of momentum space in the both denominator and the numerator, and integrating over the separation distance r in the numerator gives

$$k^{TST} = \langle J_+(\mathbf{r}^*) \rangle_R$$

$$= \frac{Q_{TS,\,\text{rot}} V(h\beta)^{-1} \int d\mathbf{r}^N \exp\{-\beta[U_{AB}(r^*) + U(\mathbf{r}^N) + V_{\text{int}}(r^*, \mathbf{r}^N)]\}}{Q_{R,\,\text{rel}} \int d\mathbf{r} \int d\mathbf{r}^N \exp\{-\beta[U_{AB}(\mathbf{r}) + U(\mathbf{r}^N) + V_{\text{int}}(\mathbf{r}, \mathbf{r}^N)]\}}$$

$$(14.13)$$

where $Q_{TS,\,\text{rot}}$ and $Q_{R,\,\text{rel}}$ are the rotational partition function of A–B at the transition state and (translational) partition function for the A–B reactants, respectively. The * indicates that we are concerned with the transition state configuration and the potentials in the numerator depend only on r since the orientation angles are integrated over in the rotational partition function (see exercises).

The integral in the denominator is the configuration integral denoted as Π

$$\Pi = \int d\mathbf{r} \int d\mathbf{r}^N \exp\{-\beta[U_{AB}(\mathbf{r}) + U(\mathbf{r}^N) + V_{\text{int}}(\mathbf{r}, \mathbf{r}^N)]\} \qquad (14.14)$$

so we rewrite Eq. (14.13) as

$$k_{TST} = \frac{1}{Q_{R,\,\text{rel}} h \beta \pi} Q_{TS,\,\text{rot}} V \exp[-\beta U_{AB}(r^*)] \int d\mathbf{r}^N$$
$$\cdot \exp\{-\beta[U(\mathbf{r}^N) + V_{\text{int}}(r^*, \mathbf{r}^N)]\} \qquad (14.15)$$

We will now introduce the radial distribution function $g(r^*)$ [12] for a given A–B distance r^* in the solvent, which is given by

$$g(r^*) = \frac{V}{\Pi} \exp\left[-\beta U_{AB}(r^*)\right] \int d\mathbf{r}^N \exp\left\{-\beta[U(\mathbf{r}^N) + V_{\text{int}}(r^*, \mathbf{r}^N)]\right\} \quad (14.16)$$

The function $g(r)$ can be thought of as the factor that, when multiplied by the bulk density, ρ, gives the local density $\rho(\mathbf{r})$ about some molecule. The term $\rho g(r)4\pi r^2\, dr$ is the probability of observing a second molecule in dr given that there is a molecule at the origin of $\mathbf{r}$.

We obtain an expression for the TST rate constant

$$k^{TST} = \frac{Q_{TS,\,\text{rot}}}{Q_{R,\,\text{rel}}h\beta} e^{-\beta E_{\text{act}}} \frac{g(r^*)}{e^{-\beta E_{\text{act}}}} \quad (14.17)$$

where $E_{\text{act}} = U_{AB}(r^*)$. The last factor in Eq. (14.17) represents the difference between the solution and gas phase rate constants.

In what follows we will attempt to understand the implications of this factor. We begin by generalizing Eq. (14.16)

$$g(r) = \frac{V}{\Pi} \int d\mathbf{r}^N\, e^{-\beta U(r,\mathbf{r}^N)} \quad (14.18)$$

and define a quantity $w(r)$ such that

$$g(r) = e^{-\beta w(r)} \quad (14.19)$$

We obtain

$$-\nabla_j w = -\frac{1}{\Pi} \int d\mathbf{r}^N \nabla_j U(r, \mathbf{r}^N)\, e^{-\beta U(r,\mathbf{r}^N)} \quad (14.20)$$

where $\nabla_j w = (\partial w/\partial x_j,\ \partial w/\partial y_j,\ \partial w/\partial x_j)$, i.e., the derivative with respect to the coordinates of the jth particle of the reacting system. In Eq. (14.20), Π is now defined as:

$$\Pi = \int d\mathbf{r}^N\, e^{-\beta U(r,\mathbf{r}^N)} \quad (14.21)$$

The left side in Eq. (14.20) expresses the force acting on particle j of the reacting system for any fixed configuration of $\mathbf{r}^N$. Therefore the right-hand side is the mean force f_j acting on molecule j (j is either A or B) averaged over the different solvent configurations, so that finally we have

$$f_j = -\nabla_j w \quad (14.22)$$

We can conclude that w is the potential that gives the mean force acting on a molecule j; it is denoted as the potential of the mean force (see Fig. 14.4). In our specific example it is the reversible work needed for bringing A and B from infinite separation to a separation r in the presence of the solvent. We represent $w(r)$ as a sum of two terms: the direct interaction between A and B ($U_{AB}(\mathbf{r}^*)$) and the indirect effective potential energy due to the solvent $\delta w(r)$. We can rewrite the TST rate constant in terms of the gas phase rate constant, k_g^{TST}, and an exponential factor involving $\delta w(r)$

$$k^{TST} = k_g^{TST}\, e^{-\beta \delta w(r^*)} \qquad (14.23)$$

where $\delta w(r^*) = w(r^*) - U_{AB}(r^*)$.

The potential of the mean force is an equilibrium quantity calculated for fixed separation distances between the reacting molecules and averaged over equilibrium distributions of solvent molecules. The solvent molecules interact with the reacting system. Calculations have to be done for each separation distance between the reacting molecules. It is by no means surprising that we are concerned with solvent molecules in equilibrium with the reacting system, this being one of the assumptions of TST. The solvent molecules do not hinder our reacting system from passing over the activation barrier; they stabilize the products so that energy is dispersed, and the products are spatially separated from each other. This ensures that recrossings will not occur.

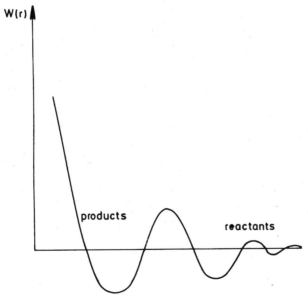

Fig. 14.4. The potential of the mean force as a function of the separation distance.

14.2 DIFFUSION INFLUENCED REACTIONS

Diffusion-controlled reactions can be written as

$$A + B \rightleftharpoons (AB)^{\#}$$

$$(AB)^{\#} \rightarrow \text{products} \qquad (14.24)$$

where the equilibrium reaction illustrates the rate of diffusion towards each other (rate constant k_1) or away from each other (rate constant k_{-1}). The two molecules A and B need to diffuse towards each other until they are close enough to react. The final reaction towards products has the rate constant k_d. Because A and B disappear upon reaction there will be a concentration gradient of B around A and we will calculate the flux and particle current of B towards A. The rate of the reaction is the current of B (number of B molecules per unit time) diffusing towards A and making an encounter with A, times the concentration of the A reactants. We define the density distribution of B around A as

$$\rho(\mathbf{r}, t) = \frac{C_B(\mathbf{r})}{C_B^*} \qquad (14.25)$$

where $C_B(\mathbf{r})$ and C_B^* are the concentration of B as a function of distance from A, and the bulk concentration of B, respectively.

As an example we consider the photostimulation of anthracene with ultrashort light pulses, producing excited singlet and triplet states of anthracene. These states are quenched by tetrabromomethane. Immediately before photostimulation the quencher is randomly distributed throughout the system, and a short time after photostimulation the quencher remains randomly distributed. About 80 years ago Smoluchowski [4] suggested that at a separation distance R (the encounter distance) between A and B the reactants react very rapidly to form products. Only when the separation distance is equal to the encounter distance do reactions occur. As boundary conditions we have

$$\rho(\mathbf{r}, 0) = 0 \quad \text{for} \quad |\mathbf{r}| < R$$
$$\rho(\mathbf{r}, 0) = 1 \quad \text{for} \quad |\mathbf{r}| \geq R$$
$$\rho(|\mathbf{r}| \rightarrow \infty, t) = 1 \quad \text{for} \quad t \geq 0$$
$$\rho(\mathbf{R}, t) = 0 \quad \text{for} \quad t \geq 0$$

We must solve the diffusion equation

$$\frac{\partial \rho(\mathbf{r}, t)}{\partial t} = D\Delta\rho(\mathbf{r}, t), \qquad (14.26)$$

where D is the diffusion constant and Δ is the Laplace operator:

$$\Delta = \frac{\partial^2}{\partial x^2} + \frac{\partial^2}{\partial y^2} + \frac{\partial^2}{\partial z^2} \tag{14.27}$$

We assume spherical symmetry, so the B molecules diffuse towards a spherically symmetric reactant A. The solution to this diffusion equation is known (see Exercise 2 in the end of this chapter), namely

$$\rho(r,t) = 1 - \frac{R}{r}\,\mathrm{erfc}\left[\frac{(r-R)}{\sqrt{4Dt}}\right] \quad (r > R) \tag{14.28}$$

where $\mathrm{erfc}(x)$ is the error function complement of the argument x.

The total number of reactant B molecules crossing a surface at r is

$$I(r) = -4\pi r^2 J = 4\pi r^2 D\,\frac{\partial C_B}{\partial r} \tag{14.29}$$

When the steady state condition is fulfilled we find that

$$\rho(r,t) = 1 - \frac{R}{r} \tag{14.30}$$

which leads to

$$C_B(r) = C_B^*\left(1 - \frac{R}{r}\right) \tag{14.31}$$

We are then able to obtain an expression for the current of B molecules towards an A molecule $I(r)$

$$I(r) = 4\pi DRC_B^* \tag{14.32}$$

This constancy reflects the steady-state concentration of B around A. Finally we obtain

$$-\frac{dC_A}{dt} = IC_A \tag{14.33}$$

where C_A is the concentration of A molecules. We can identify the rate constant for the reaction in the steady-state limit

$$k(\infty) = 4\pi DR \tag{14.34}$$

If we do not make the steady-state approximation we find that the current $I(r)$ is given by

$$I(r) = 4\pi DR \left[1 + \frac{R}{\sqrt{\pi Dt}} \right] C_B^* \tag{14.35}$$

and by using Eq. (14.33) we identify the rate constant as

$$k(t) = 4\pi DR \left[1 + \frac{R}{\sqrt{\pi Dt}} \right] \tag{14.36}$$

In order to obtain rate constant units like dm^3/mol we have to multiply Eq. (14.36) by $10^3 N_A$ where N_A is Avogadro's number. The result in Eq. (14.36) gives, for t approaching infinity, the rate constant in the steady-state limit. The time dependence of the Smoluchowski rate coefficient [4] is due to the transient concentration of B in excess of the steady-state concentration.

14.3   THE DIFFUSION MODEL AND CHEMICAL ACTIVATION

Next we will consider how one can incorporate chemical rates into the diffusion model. This will be important when the activation process is comparable to or slower than the rate of approach of reactants in the formation of encounter pairs. In this case it is no longer satisfactory to say that the reactants cannot coexist within a distance R of one another. Because the rate of reaction of the activation process is finite, we obtain a lifetime of the encounter that is nonzero. In the 1930s, Collins and Kimball [13] introduced the following model: the chemically activated process, which leads to the products from the encounter pair, occurs at a rate proportional to the probability that the encounter exists. The encounter pair is defined as a pair of reactants that lie within a distance of R to $R + dR$ of one another and the probability that B lies within this range of A is $\rho(R)$. The rate of reaction for encounter pairs is $k_{act}\rho(R)$ where k_{act} is the second-order rate coefficient for the reaction of A and B when they are nearly in contact and close enough to react with each other. The rate at which B diffuses towards A is known and the rate of reaction for the encounter pair is $k_{act}\rho(R)$. At steady state they are equal

$$k_{act}\rho(R) = 4\pi R^2 D \left(\frac{\partial \rho}{\partial r} \right)_{r=R} \tag{14.37}$$

This is called the partially reflecting boundary condition, and our set of boundary conditions becomes

$$\rho(\mathbf{r}, 0) = 0 \qquad \text{for} \quad |\mathbf{r}| < R$$

$$\rho(\mathbf{r}, 0) = 1 \qquad \text{for} \quad |\mathbf{r}| \geq R$$

$$\rho(|\mathbf{r}| \to \infty, t) = 1 \qquad \text{for} \quad t \geq 0$$

$$k_{act}\rho(R, t) = 4\pi R^2 D \left(\frac{\partial \rho}{\partial r} \right)_{r=R} \qquad \text{for} \quad t \geq 0$$

The diffusion equation is solved by performing a Laplace transformation. We obtain in Laplace space the following solution as

$$\underline{\rho}(r, s) = \frac{1}{s} + \frac{A}{r} \exp\left[r\left(\frac{s}{D} \right)^{1/2} \right] + \frac{B}{r} \exp\left[-r\left(\frac{s}{D} \right)^{1/2} \right] \qquad (14.38)$$

$\underline{\rho}(r, s)$ is the Laplace transform (see appendix C). In order to make use of the boundary condition in Eq. (14.38) we must find its Laplace transform and thus obtain

$$k_{act} \underline{\rho}(R, s) = 4\pi R^2 D \left(\frac{\partial \underline{\rho}}{\partial r} \right)_{r=R} \qquad (14.39)$$

By making use of the boundary conditions in Eqs. (14.38) and (14.39) we obtain

$$A = 0 \qquad (14.40)$$

and

$$B = -\frac{R}{s} \exp\left[R\left(\frac{s}{D} \right)^{1/2} \right] \frac{k_{act}}{k_{act} + 4\pi RD + 4\pi R^2 \sqrt{sD}} \qquad (14.41)$$

We can invert the Laplace transform and obtain:

$$\rho(r,t) = 1 - \frac{R}{r} \frac{k_{act}}{k_{act} + 4\pi RD} \left\{ erfc\left[\frac{(r-R)}{\sqrt{4Dt}}\right] \right.$$

$$- \exp\left[(k_{act} + 4\pi RD)\frac{(r-R)}{4\pi R^2 D}\right] \exp\left[(k_{act} + 4\pi RD)^2 \frac{t}{16\pi^2 R^4 D}\right]$$

$$\left. \cdot erfc\left[\frac{(r-R)}{\sqrt{4Dt}} + \frac{(k_{act} + 4\pi RD)}{4\pi R^2 \sqrt{\frac{D}{t}}}\right] \right\} \qquad (14.42)$$

For the case where the rate of the activation process goes to infinity we obtain

$$\rho(r,t) = 1 - \frac{R}{r} erfc\left[\frac{r-R}{\sqrt{4Dt}}\right] \qquad (14.43)$$

which of course is the same as Eq. (14.28). In the steady-state limit, that is for $t \to \infty$, we have from Eq. (14.42)

$$\rho(r, t \to \infty) = 1 - \frac{R}{r} \frac{k_{act}}{k_{act} + 4\pi RD} \qquad (14.44)$$

Presently, we are able to determine the rate coefficient

$$k(t) = k_{act}\rho(R,t) = \frac{4\pi RD k_{act}}{k_{act} + 4\pi RD}$$

$$\cdot \left\{ 1 + \frac{k_{act}}{4\pi RD} \exp\left[\frac{Dt}{R^2}\left(1 + \frac{k_{act}}{4\pi RD}\right)^2\right] erfc\left[\sqrt{\frac{Dt}{R^2}}\left(1 + \frac{k_{act}}{4\pi RD}\right)\right] \right\}, \qquad (14.45)$$

which is a rather complicated function. For times accessible in standard experiments the arguments in the erfc and exp functions are large and we have

$$k(t) = \frac{4\pi RD k_{act}}{k_{act} + 4\pi RD} \left[1 + \frac{k_{act}}{k_{act} + 4\pi RD}\left(\frac{R}{\sqrt{\pi Dt}}\right)\right] \qquad (14.46)$$

In the steady state limit the rate constant is expressed as the following

$$\frac{1}{k(t \to \infty)} = \frac{1}{4\pi RD} + \frac{1}{k_{act}} \qquad (14.47)$$

which is an expression one encounters very often in physics and chemistry when branching (more than transport route) of particle transport is involved.

REFERENCES

[1] H. Eyring, J. Chem. Phys. **3**,107(1935).

[2] M. G. Evans and M. Polanyi, Trans. Faraday Soc. **34**,11(1938).

[3] R. A. Ogg and M. Polanyi, Trans. Faraday Soc. **31**,604(1935).

[4] M. Smoluchowski, Z. Phys. Chem. **92**,129(1918).

[5] P. Debye, Trans. Faraday Soc. **82**,265(1942).

[6] L. Onsager, Phys. Rev. **54**,554(1938).

[7] H. A. Kramers, Physica **7**,284(1940).

[8] R. A. Marcus, J. Chem. Phys. **24**,966,979(1956); **43**,679(1965).

[9] J. Franck and E. Rabinowitch, Trans. Faraday Soc. **30**,120(1934).

[10] R. F. Grote and J. T. Hynes, J. Chem. Phys. **73**,2715(1980).

[11] M. Born, Z. Phys. **1**,45(1920).

[12] D. A. McQuarrie, "Statistical Mechanics," Harper & Row, New York, 1976.

[13] F. C. Collins and G. E. Kimball, J. Colloid. Sci. **4**,425(1949).

EXERCISES

1. Derive Eq. (14.15) from Eq. (14.12).

2. Derive Eq. (14.28).

3. Two molecules A and B react with the rate constant k_{act} at a distance $R = 4$ Å. The diffusion constant D_{AB} is 10^{-5} cm^2/sec. Find the rate constant k_d if the reaction is diffusion controlled. Discuss on what time scale the steady-state limit is obtained. Find an expression in terms of k_d and k_{act} for the rate coefficient k (or $1/k$) in the steady-state limit and calculate k if $k_{act} = 10^{-11}$ cm^3/sec.

4. Assume that the reaction

$$A + B \to AB$$

is diffusion controlled and calculate the rate constant at room temperature in a solution of ether, ethanol, and glycerol. The viscosity η for ether is $2.33 \cdot 10^{-4}$ kg m^{-1} sec^{-1}, for ethanol is $11.9 \cdot 10^{-4}$ kg m^{-1} sec^{-1}, and for glycerol is 1.49 kg m^{-1} sec^{-1}. Use $D_A \sim D_B$, $D_{AB} = D_A + D_B$; the molecular radius for both A and B is $a = 3$ Å, and $R = 2a$. In which of the three solutions is the reaction most likely diffusion controlled?

5. The reaction

$$H_3O^+ + OH^- \rightarrow 2H_2O \tag{14.48}$$

is diffusion controlled; the effective diffusion constant $D = D_{H_3O^+} + D_{OH^-} = 14.6 \cdot 10^{-5}$ cm^2/sec and the reaction radius $R = 7.5$ Å.

a. Give an expression for the rate constant and calculate the value.

b. Experimentally one finds $k(T) = 2.16 \pm 0.33 \ 10^{-10}$ cm^3/sec at 298 K. The deviation from the result above is due to the neglect of the migration term, i.e., the flux of A molecules towards B molecules should be:

$$J_A = -D \frac{\partial C_A}{\partial r} - \frac{z_A}{k_B T} C_A D \nabla \phi \tag{14.49}$$

where ϕ is

$$\phi = \frac{z_B e^2}{\epsilon r} \tag{14.50}$$

and z_a and z_b are the charges on the two ions, ϵ the dielectric constant of the solvent, and D the diffusion constant. The flux over a sphere with radius r is now

$$I(r) = 4\pi r^2 D \left[\frac{\partial C_A(r)}{\partial r} + \frac{C_A(r)}{k_B T} \frac{\partial U(r)}{\partial r} \right] \tag{14.51}$$

where

$$U(r) = \frac{z_A z_B e^2}{\epsilon r} \tag{14.52}$$

c. Show that the expression

$$C_A(r) = C_A^* \frac{[(exp(-\delta/R) - exp(-\delta/r)]}{exp(-\delta/R) - 1} \tag{14.53}$$

satisfies the boundary conditions and find the value of δ for which $I(r)$ = constant. Investigate the limit $\delta \rightarrow 0$

d. Use $k(T) = I(R,T)/C_A^*$ to calculate the rate constant at a temperature of 298 K. The dielectric constant of water is 80.

6. In the example at the end of Chapter 12 we assumed that the transfer of an electron between the electrode and the molecule was infinitely fast. In this exercise we consider the case where the kinetics of the electron transfer is dependent on the potential of the electrode.

$$P + e \rightarrow Q \qquad (14.54)$$

The electron transfer rate constant is k_{ET}. The planar electrode is positioned on the x-axis at $x = 0$ and is turned on at time $t = 0$. The compound P has a diffusion coefficient D and a bulk concentration C^*. The area of the electrode is A.

a. Establish the diffusion equation for the compound P along with the appropriate boundary conditions and an expression for the current. We assume that the electron transfer rate constant is given by

$$k_{ET} = k_{ET}^o e^{bE} \qquad (14.55)$$

where $b = \beta F/RT$. The electric potential of the electrode is given by E. The three variables F, R, and T are Faraday's constant, the gas constant, and the temperature, respectively. For $E = 0$ the rate constant is k_{ET}^o.

b. Determine the Laplace transform of the flux of the compound P at the surface.

c. Combine the result from (**b**) with the boundary conditions in order to obtain an expression for the surface concentration of P, and calculate the current.

d. Plot the current function $G = i/FAk_{ET}C^*$ versus time for different values of k_{ET}: $1 \cdot 10^{-5}$ m/sec; $5 \cdot 10^{-5}$ m/sec; $1 \cdot 10^{-4}$ m/sec.

15

KRAMERS' THEORY

There are two key assumptions that underlie many-body rate theory:

   (i) Thermodynamic equilibrium prevails throughout the entire system for all
       the degrees of freedom. Any nonequilibrium effects that result in pertur-
       bations of the Boltzmann distribution are neglected.
  (ii) Correlated returns (recrossings) of a system that has crossed the
       (parabolic) bounding hypersurface through a saddle point are neglected.

This chapter considers an approach for going beyond these key assumptions,
and the problem can be depicted in the following way: A particle moves in an
external field of force but in addition to this is subjected to the irregular forces
of a surrounding medium possessing an equilibrium temperature. This was dealt
with in Section 12.2 on Brownian motion.

The conditions are such that the particle is originally confined in a potential
hole but may escape in the course of time by passing over a potential barrier.
The quantity that we wish to calculate is the probability of escape as a function
of temperature and viscosity of the medium; we then wish to compare this with
the results from TST [1]. First we construct and discuss the equation of motion
or diffusion obeyed by a density distribution of particles in phase space. We
will obtain analytical results for limiting cases.

The principles of Brownian motion in phase space can be outlined in terms of
a reaction system represented by an effective particle of mass M crossing a one-
dimensional potential barrier $U(x)$ (see Fig. 15.1) and a solvent whose influence
on the reacting system is given as a Langevin equation. As was seen in Section
12.2 on Brownian motion (Eq. (12.34)) we can write the Langevin equation

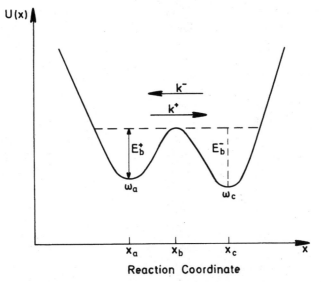

Fig. 15.1. The potential $U(x)$ used in Kramers' theory.

$$M\frac{dv}{dt} = F_{\text{ext}}(t) - \gamma Mv + F(t), \qquad (15.1)$$

where the first term on the right-hand side is the external force, the second term is the linear damping term, and the last term the fluctuating force. The reacting system is represented by the mass M, the center of mass coordinate x, and the velocity v. The external force is given as

$$F_{\text{ext}}(t) = -\frac{\partial U}{\partial x} \qquad (15.2)$$

The fluctuating force $F(t)$ has the following properties

$$\langle F(t) \rangle = 0 \qquad (15.3)$$
$$\langle F(t)F(t') \rangle = 2M\gamma kT\delta(t - t'), \qquad (15.4)$$

where the angle brackets denote ensemble averages, and

$$\int_{-\infty}^{\infty} \langle F(t)F(t') \rangle \, d(t - t') = 2M\gamma kT \qquad (15.5)$$

where Eq. (15.5) gives the definition of γ. Kramers' model [1] for a chemical reaction consists of a classical particle of mass M moving in a one-dimensional asymmetric double-well potential $U(x)$. The particle coordinate x corresponds to the reaction coordinate. The minima of the potential U at x_a and x_c denote the reactant and the product states, respectively (see Fig 15.1). The maximum of $U(x)$ is located at x_b, the barrier separating the two states and corresponds to the location of the transition state. All the remaining degrees of freedom for both reacting and solvent molecules constitute a heat bath at a temperature T, whose total effect on the reacting particle is described by a fluctuating force $F(t)$ and by a linear damping force $-\gamma Mv$, where γ is a constant damping rate.

The resulting two-dimensional stochastic dynamics for the reaction coordinate x and the velocity v is called Markovian and the time evolution of the probability density $P(x,v,t)$ is governed by the Klein-Kramers' equation [2,3]

$$\frac{\partial P(x,v,t)}{\partial t} = \left[-\frac{\partial}{\partial x}v + \frac{\partial}{\partial v}\left(\frac{\partial U}{M\partial x} + \gamma v \right) + \frac{\gamma kT}{M}\frac{\partial^2}{\partial v^2} \right] P(x,v,t), \quad (15.6)$$

which can be rewritten to give

$$\frac{\partial P(x,v,t)}{\partial t} + v\frac{\partial}{\partial x}P(x,v,t) - \frac{1}{M}\frac{\partial}{\partial v}\left[\left(\frac{\partial U}{\partial x} \right) P(x,v,t) \right]$$
$$= \gamma\frac{\partial}{\partial v}(vP(x,v,t)) + \frac{\gamma kT}{M}\frac{\partial^2 P(x,v,t)}{\partial v^2} \quad (15.7)$$

15.1   THE KLEIN-KRAMERS' EQUATION

Let us consider how this equation is obtained. For this purpose we will consider how the density distribution function $P(x,v,t)$ in phase space will evolve in time [2,3]. We begin with a time interval Δt, which is long compared to the periods of fluctuation of random solvent forces, but short compared to intervals during which the position and velocity of the Brownian particle changes by appreciable amounts. Under these circumstances we should be able to derive $P(x,v,t)$. The diffusion equation will give us the following changes in x and v:

$$\Delta x = v\Delta t \quad (15.8)$$

$$M\Delta v = -\frac{\partial U}{\partial x}\Delta t - \gamma Mv\Delta t + F(t)\Delta t \quad (15.9)$$

where we define the last term in Eq. (15.9) to be

$$B(\Delta t) = F(t)\Delta t, \quad (15.10)$$

the net acceleration arising from fluctuations that occur in a time dt. The change in velocity is then

$$\Delta v = -M^{-1}\frac{\partial U}{\partial x}\Delta t - \gamma v \Delta t + B(\Delta t)M^{-1} \tag{15.11}$$

We assume that we know the function P at time t, and would like to determine the function P at later time $t + \Delta t$. At t we can write the position as $x - \Delta x$ and the velocity as $v - \Delta v$, while at $t + \Delta t$ the position of the particle is then x and its velocity v. We have the corresponding functions: $P(x - \Delta x, v - \Delta v, t)$ and $P(x, v, t + \Delta t)$ at time t and at later time $t + \Delta t$. The distribution function $P(x, v, t + \Delta t)$ is then

$$P(x, v, t + \Delta t) = \int d(\Delta x) \int d(\Delta v)P(x - \Delta x, v - \Delta v, t)$$
$$\cdot \Psi(x - \Delta x, v - \Delta v; \Delta x, \Delta v) \tag{15.12}$$

where $\Psi(x - \Delta x, v - \Delta v; \Delta x, \Delta v)$ is the transition probability in x, v space and is a function that answers the question: What is the probability for getting to a point in phase space (x, v) from another point in phase space $(x - \Delta x, v - \Delta v)$? Since Eq. (15.8) holds we can express Ψ in terms of a transition probability in velocity space Φ and we write

$$\Psi(x - \Delta x, v - \Delta v; \Delta x, \Delta v) = \delta(\Delta x - v\Delta t)\Phi(x - \Delta x, v - \Delta v; \Delta v) \tag{15.13}$$

Integrating Eq. (15.12) with respect to $d(\Delta x)$ we have

$$P(x, v, t + \Delta t) = \int d(\Delta v)P(x - \Delta x, v - \Delta v, t)\Phi(x - \Delta x, v - \Delta v; \Delta v), \tag{15.14}$$

which can be rewritten as

$$P(x + v\Delta t, v, t + \Delta t) = \int d(\Delta v)P(x, v - \Delta v, t)\Phi(x, v - \Delta v; \Delta v) \tag{15.15}$$

In order to proceed we expand in a Taylor series around x, v, t; thus for the left-hand side (lhs) of Eq. (15.15) we obtain

$$\text{lhs} = P(x, v, t) + \frac{\partial P}{\partial t}(\Delta t) + \frac{\partial P}{\partial x}(v\Delta t)$$
$$= P(x, v, t) + \left[\frac{\partial P}{\partial t} + v\frac{\partial P}{\partial x}\right]\Delta t \tag{15.16}$$

For the right-hand side (rhs) of Eq. (15.15) we obtain

$$\text{rhs} = \int d(\Delta v) \left[P(x,v,t) + \frac{\partial P}{\partial v}(-\Delta v) + \frac{1}{2}\frac{\partial^2 P}{\partial^2 v}(-\Delta v)^2 \right]$$
$$\cdot \left[\Phi(x,v;\Delta v) + \frac{\partial \Phi}{\partial v}(-\Delta v) + \frac{1}{2}\frac{\partial^2 \Phi}{\partial v^2}(-\Delta v)^2 \right] \qquad (15.17)$$

Finally we have

$$\left[\frac{\partial P}{\partial t} + v\frac{\partial P}{\partial x} \right]\Delta t = \int d(\Delta v)\left[\frac{\partial P}{\partial v}(-\Delta v)\Phi(x,v;\Delta v) + P(x,v,t)\frac{\partial \Phi}{\partial v}(-\Delta v) \right]$$
$$+ \int d(\Delta v)\left[\frac{\partial P}{\partial v}\frac{\partial \Phi}{\partial v}(\Delta v)^2 + \frac{1}{2}\frac{\partial^2 P}{\partial v^2}\Phi(x,v;\Delta v)(\Delta v)^2 \right.$$
$$\left. + \frac{1}{2}\frac{\partial^2 \Phi}{\partial v^2}P(x,v,t)(\Delta v)^2 \right] \qquad (15.18)$$

and define

$$\langle \Delta v \rangle = \int d(\Delta v)\Delta v \Phi(x,v;\Delta v) \qquad (15.19)$$

$$\langle (\Delta v)^2 \rangle = \int d(\Delta v)(\Delta v)^2 \Phi(x,v;\Delta v) \qquad (15.20)$$

Using these definitions we obtain

$$\left[\frac{\partial P}{\partial t} + v\frac{\partial P}{\partial x} \right]\Delta t = -\frac{\partial}{\partial v}(P\langle \Delta v \rangle) + \frac{1}{2}\frac{\partial^2}{\partial v^2}[P\langle (\Delta v)^2 \rangle], \qquad (15.21)$$

and from Eqs. (15.4) and (15.11) (see also Exercise 1)

$$\langle \Delta v \rangle = -\left[\gamma v + \frac{1}{M}\left(\frac{\partial U}{\partial x} \right) \right]\Delta t \qquad (15.22)$$

$$\langle (\Delta v)^2 \rangle = \left(2\frac{\gamma k T}{M} \right)\Delta t, \qquad (15.23)$$

which results in

$$\left[\frac{\partial P}{\partial t} + v\frac{\partial P}{\partial x}\right] = \frac{\partial}{\partial v}\left\{\left[\gamma v + \frac{1}{M}\left(\frac{\partial U}{\partial x}\right)\right]P\right\} + \left(\frac{\gamma kT}{M}\right)\frac{\partial^2}{\partial v^2}(P), \quad (15.24)$$

which when rewritten gives

$$\frac{\partial P}{\partial t} + v\frac{\partial P}{\partial x} - \frac{1}{M}\frac{\partial U}{\partial x}\frac{\partial P}{\partial v} = \gamma\frac{\partial}{\partial v}(vP) + \frac{\gamma kT}{M}\frac{\partial^2 P}{\partial v^2} \qquad (15.25)$$

Eqs. (15.24) and (15.25) are both the Klein-Kramers' equation.

15.2   CHEMICAL ACTIVATION

Let us consider the qualitative behavior of the problem we are investigating in the case of $kT \ll$ the activation energy. In this case the random force acts only as a small perturbation so our Langevin equation becomes

$$\frac{dv}{dt} = -\frac{1}{M}\frac{\partial U}{\partial x} - \gamma v \qquad (15.26)$$

Hence the reaction coordinate will relax toward one of the minima of the potential, where the system will stay for an exceedingly long time until eventually the accumulated action of the random forces will drive it over the barrier into a neighboring state. If on the other hand, the thermal energy kT is comparable to or even larger than the activation energy, then the particle can move almost freely from x_a to x_c (Fig. 15.1 or 15.2). A rate description makes no sense in this case; therefore, we will only consider cases where the activation energy is much larger than kT. The strength of the interaction between the reaction coordinate and the remaining degrees of freedom is fixed by a single constant, the damping rate γ. We will consider two limiting cases: (i) strong friction (ii) low friction. The evaluation of the rate requires one additional assumption: thermal equilibrium in the initial well is maintained at all times and the distribution function is given by a Boltzmann distribution probability

$$P(x, v) = Q^{-1}\exp\left\{-\beta\left[\frac{Mv^2}{2} + U(x)\right]\right\} \qquad (15.27)$$

where Q denotes the partition function and $\beta = 1/kT$. We determine the steady state escape rate say from A $\rightarrow$ C by considering a stationary situation in which a steady probability current from A to C is maintained by sources and sinks. The sources supply the A-well with particles at energies that are a few kT below the top of the barrier. These particles are thermalized before they eventually leave

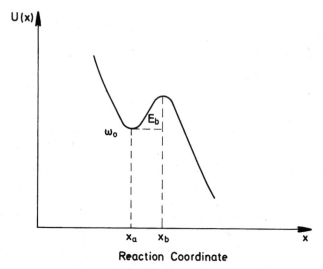

Fig. 15.2. The bistable potential as a function of reaction coordinate.

the well over the barrier. Beyond the barrier the particles are removed again by sinks. The total probability flux over the barrier is then given as

$$j = k_{AC} n_{AC} \tag{15.28}$$

where n_{AC} is the population of the A-well and k_{AC} is the rate for A $\rightarrow$ C. We have the following potential around the barrier top, where x is confined to a small neighborhood around the parabolic top

$$U(x) = U(x_b) - \tfrac{1}{2} M \omega_b^2 (x - x_b)^2 \tag{15.29}$$

Near the bottom of the A-well, all particles are thermalized and $P(x, v)$ is given by Eq. (15.27). On the other hand, near the bottom of the C-well all particles are removed by the sinks; $P(x, v) = 0$ for $x \approx x_c > x_b$. If we have $P(x, v)$ we can find the population of the A-well

$$n_a = \int_{A\text{-well}} dx \, dv \, P(x, v), \tag{15.30}$$

and the flux over the barrier

$$j = \int_{-\infty}^{\infty} dv \, v P(x_b, v) \tag{15.31}$$

The probability function $P(x, v)$ around the barrier obeys the stationary equation, Eq. (15.25), i.e., $\partial P/\partial t = 0$ and we obtain the following equation for $P(x, v)$, i.e.,

$$0 = \left\{ -v \frac{\partial}{\partial x} - \frac{\partial}{\partial v} [\omega_b^2(x - x_b) - \gamma v] + \frac{\gamma kT}{M} \frac{\partial^2}{\partial v^2} \right\} P(x, v) \qquad (15.32)$$

where we have introduced $\partial U/\partial x$ using Eq. (15.29). We must find a solution to Eq. (15.32) subject to the conditions

$$P(x, v) = Q^{-1} \exp\left[-\frac{1}{kT} \left(\frac{Mv^2}{2} + U(x) \right) \right] \qquad \text{for } x \approx x_a \qquad (15.33)$$

$$P(x, v) = 0 \qquad \text{for } x \approx x_c > x_b \qquad (15.34)$$

In order to find $P(x, v)$ we make the Ansatz (assumption) that

$$P(x, v) = Q^{-1} Y(x, v) \exp\left[-\frac{1}{kT} \left(\frac{Mv^2}{2} + U(x) \right) \right] \qquad (15.35)$$

where $Y(x, v)$ is to be identified below. For $x \approx x_b$ we use the stationary equation from Eq. (15.32) to find $Y(x, v)$, resulting in

$$0 = \left\{ -v \frac{\partial}{\partial x} - [\omega_b^2(x - x_b) + \gamma v] \frac{\partial}{\partial v} + \left(\frac{\gamma kT}{M} \right) \frac{\partial^2}{\partial v^2} \right\} Y(x, v) \qquad (15.36)$$

subject to the conditions: $Y(x, v) = 1$ for $x \approx x_a$, and $Y(x, v) = 0$ for $x \approx x_c$. Such behavior of Y may be achieved if Y is a function of a variable $u = v - a(x - x_b)$, i.e., a linear combination of x and v. Kramers solved this problem and found

$$Y(\tilde{u}) = K \int_{u_c}^{\tilde{u}} du \exp\left(-\frac{\lambda_+ M u^2}{2\gamma kT} \right) \qquad (15.37)$$

where

$$\lambda_+ = -\frac{\gamma}{2} + \sqrt{\omega_b^2 + \left(\frac{\gamma}{2} \right)^2} \qquad (15.38)$$

and K is a normalization constant determined such that $Y[v - a(x_a - x_b)] \sim 1$, $u_c = v - a(x_c - x_b)$ and $a = \gamma + \lambda_+$. The well population is given by

$$n_a = \frac{K}{Q} \frac{2\pi k T}{M\omega_0} \sqrt{\frac{2\pi\gamma k T}{\lambda_+ M}} \exp\left[-\beta U(x_a)\right] \tag{15.39}$$

where we have expanded the potential around x_a

$$U(x) = U(x_a) + \tfrac{1}{2} M\omega_o^2 (x - x_a)^2 \tag{15.40}$$

and where ω_o is the frequency for the motion at the bottom of the A-well. The flux over the barrier is given by

$$j = \frac{1}{Q} K \frac{kT}{M} \sqrt{\frac{2\pi k T \gamma}{M a}} \exp\left[-\beta U(x_b)\right] \tag{15.41}$$

and the rate is given by j/n_a, i.e.,

$$k_{AC} = \frac{-\frac{\gamma}{2} + \sqrt{\omega_b^2 + \frac{\gamma^2}{4}}}{\omega_b} \frac{\omega_o}{2\pi} \exp\left(-\frac{E_b}{kT}\right)$$
$$= k^{Kr} k^{TST} \tag{15.42}$$

where Kr denotes that this expression was obtained by Kramers and where

$$E_b = U(x_b) - U(x_a), \tag{15.43}$$

which is the activation barrier.

For convenience we rewrite the Kramers' correction to the transition state result as

$$k^{Kr} = -\frac{\gamma}{2\omega_b} + \sqrt{1 + \left(\frac{\gamma}{2\omega_b}\right)^2} \tag{15.44}$$

For a solvent with a finite value of γ we obtain a Kramers' correction that is less than one. If the ratio of γ/ω_b is small, k^{Kr} approaches one, the solvent friction is ineffective in inducing recrossings, the reaction is a free flight passage from A to C and hence $k_{AC} = k^{TST}$. If the ratio γ/ω_b is much larger than one (the overdamped region) the Kramers' correction approaches ω_b/γ and for high friction

$$k_{AC} = \frac{\omega_b}{\gamma} \frac{\omega_o}{2\pi} \exp\left(-\frac{E_b}{kT}\right), \qquad (15.45)$$

which shows that for γ approaching infinity the rate approaches zero. Here the solvent is able to induce recrossings. Introducing the diffusion constant, D, we express the rate as

$$k_{AC} = k^{TST} \frac{\omega_b M}{kT} D \qquad (15.46)$$

illustrating the diffusive behavior.

We have considered Kramers' approach for moderate-to-high friction where the associated nonequilibrium effects can be modelled in terms of diffusional surface recrossings over the barrier.

Let us now consider the second type of nonequilibrium effect mentioned in the beginning of this section: that related to the deviation from thermal equilibrium inside the initial well. Such effects play an increasingly important role for reactions where the solvent has very little friction, since in these situations the particle or reacting system suffers infrequent collisions. The energy will be the only slowly relaxing variable. In this regime Kramers assumed that the number of collisions per unit time would be sufficiently small that the rate of energy transfer would be slow compared to an oscillation frequency ω_o in well A. One can rewrite Eq. (15.25) so it is transformed from phase space to energy space and obtain [1]

$$\frac{\partial P(E,t)}{\partial t} = \frac{\partial}{\partial E}\left[D(E)\left(\frac{\partial}{\partial E} + \frac{1}{kT}\right)\omega(E)P(E,t)\right] \qquad (15.47)$$

where $D(E)$ is a diffusion coefficient.

We define the probability flux in energy space, $I(E)$, as the area inside a curve of constant energy, i.e.,

$$I(E) = \oint p\,dq = M \oint v\,dx \qquad (15.48)$$

and $\omega(E) = \partial E/\partial I$. This allows us to calculate $P(E,t)\,dI(E)$, that is the fraction of the ensemble lying inside the ring-shaped area $dI(E)$. Letting the friction γ take on a value where it is much smaller than $kT/I(E)$ we find that

$$D(E) = \omega(E)D(I) = kT\gamma I(E) \qquad (15.49)$$

The potential is the one used previously. As soon as the reacting system has acquired an energy slightly larger than the activation barrier E_b, the Brownian particle escapes the A-well. The rate is given by the probability flux of particles in energy space through E_b, divided by the population of the A-well.

We will now find the stationary probability distribution using the assumptions that for x larger than x_b the probability distribution $P(E,t)$ is zero and that the energy barrier is much larger than kT. We obtain a rate constant of

$$k_{AC} = \frac{\gamma I(E)}{kT} \frac{\omega_o}{2\pi} e^{-E_b/kT} \tag{15.50}$$

We can make use of the definition in Eq. (15.49) of the diffusion constant and rewrite this expression as

$$k_{AC} = \frac{D(E)}{(kT)^2} \frac{\omega_o}{2\pi} e^{-E_b/kT} \tag{15.51}$$

we then note that for solvents in this limit of low friction we obtain a rate that is proportional to the friction γ.

Kramers' theory predicts that for solvents with low friction the rate increases for increasing γ, whereas for solvents with high friction the rate decreases for increasing γ.

REFERENCES

[1] H. A. Kramers, Physica **7**,284(1940).

[2] N. Wax, ed., "Selected Papers on Noise and Stochastic Processes," Dover, New York, 1954.

[3] W. Coffey, Adv. Chem. Phys., **LXIII**,69(1985).

SUGGESTED READING

P. Hänggi, P. Talkner and M. Borkovec, "Reaction-rate theory: fifty years after Kramers," Rev. Mod. Phys. **62**,251(1990).

EXERCISES

1. Use Eq. (15.5) to derive Eq. (15.23).

2. Derive Eq. (15.37) using Eq. (15.36). Find K when $u_c \sim -\infty$ and $u_a = v - a(x_a - x_b) \sim \infty$.

3. Derive Eq. (15.39) and insert the value for K found above.

4. Derive Eq. (15.41). Use

$$\int_0^\infty dx \exp(-\mu x^2) x \int_0^x dy \exp(-\beta y^2) = \frac{\sqrt{\pi}}{4} \frac{1}{\mu \sqrt{\mu + \beta}} \tag{15.52}$$

5. The potential along the reaction path for a proton transfer reaction is given as

$$V(x) = -\tfrac{1}{2}a_0 x^2 + \tfrac{1}{4}c_0 x^4 \tag{15.53}$$

where $a_0 = 415 \text{ Å}^{-2} \text{ kJ mol}^{-1}$ and $c_0 = 1636 \text{ Å}^{-4} \text{ kJ mol}^{-1}$. Find ω_b and ω_0. Assume that $D = 10^{-4} \text{ cm}^2/\text{sec}$ and use the relation $D = kT/M\gamma$ to find k^{Kr} at 300 K. Finally determine the rate constant at this temperature.

6. For chemical reactions in liquids one finds the following expression for the rate constant:

$$k_{TST} = \frac{1}{2\pi} \frac{\displaystyle\prod_{i=0}^{N} \omega_i^{(0)}}{\displaystyle\prod_{i=1}^{N} \omega_i^{(b)}} \exp(-\beta E_b) \tag{15.54}$$

Explain how this expression can be obtained from transition state theory. What assumption must be made in order to reduce this expression to the transition state limit of Kramers' theory? According to Kramers' theory the rate constant can be written as:

$$k(T) = \frac{\kappa \gamma}{\omega_b} k_{TST} \tag{15.55}$$

Discuss the two limits for κ.

7. For a proton transfer reaction one finds experimentally that $k_{TST} = 3.1 \cdot 10^9$ sec^{-1} at 298 K. Assume that the potential for the reaction is:

$$U(x) = \tfrac{1}{2}U_0[1 + \cos(2\pi x/L)] \, (-L < x < L) \tag{15.56}$$

with the barrier $U_0 = 20 \text{ kJ/mol}$. Using Kramers' theory, calculate the value of L corresponding to the experimental observation. In a different solvent one finds that $k(T) = 1.0 \cdot 10^9 \text{ sec}^{-1}$. Determine γ and the diffusion constant D at 298 K. The potential parameters are assumed to be the same as before.

8. Consider the Klein-Kramers' equation for two cases:
   a. A constant force
   b. The potential is given by a simple harmonic oscillator potential.

16

THE CLASSICAL MODEL OF ELECTRON TRANSFER REACTIONS IN SOLUTION

16.1 INTRODUCTION

In this chapter we will consider a very important class of chemical reactions in solution, namely electron transfer (ET) reactions. We will consider the classical approach of R. A. Marcus [1]. In this model we make the following basic assumptions:

(1) Every single state (reactant, product or transition state) that is assumed to be involved in ET can be described by classical electrostatic theory.

(2) No explicit reference to quantum states is made for either the solvent or the reacting compounds.

(3) The system states are defined by mechanistic concepts and their energies by a model of point or spherical charges contained within cavities immersed in the solvent.

(4) The medium is represented as continuous and enters the description of the energetics of the system through the polarizability and dielectric constant for the medium.

(5) The reaction rate depends on the population of the mechanistically defined states, and the states are assumed to be governed by a Boltzmann distribution.

The ET reaction takes place between two molecules A and B with the charges e_1^* and e_2^*, respectively. They are initially an infinite distance from each other. This defines the initial state I. The two molecules approach each

136

other by diffusion, and at some point they form a precursor or encounter complex A–B. The next step involves reorganization of the precursor complex to a configuration appropriate for ET. Electron transfer then takes place within the reorganized precursor complex $[A|B]^*$ yielding the reorganized successor complex $[A^+|B^-]^*$. We obtain a doubly degenerate transition state ($[A|B]^*$ $[A^+|B^-]^*$) differing only in the localization of the electron that will be transferred. These two states differ only with respect to charge distribution; the charges for the two molecules in the $[A|B]^*$ state are e_1^* and e_2^*, respectively, and the charges for the two molecules in the $[A^+|B^-]^*$ state are e_1 and e_2, respectively. Vibrational relaxation of the reorganized successor complex and its dissociation yields the separated products A^+ and B^-. Schematically we can write this as:

$$A + B \underset{k_{-1}}{\overset{k_1}{\rightleftharpoons}} [A|B]^*$$

$$[A|B]^* \underset{k_{-2}}{\overset{k_2}{\rightleftharpoons}} [A^+|B^-]^*$$

$$[A^+|B^-]^* \overset{k_3}{\rightarrow} A^+ + B^- \tag{16.1}$$

The rate of formation(s) of products is

$$S = k_3[[A^+|B^-]^*] \tag{16.2}$$

and assuming that the concentrations of $[A|B]^*$ and $[A^+|B^-]^*$ are time-independent we have, using the steady-state approximation,

$$S = k_{obs}[A][B] \tag{16.3}$$

where

$$k_{obs} = \frac{k_1}{1 + (1 + k_{-2}/k_3)(k_{-1}/k_2)} \tag{16.4}$$

and the rate is controlled by the rate of preparation of the state $[A|B]^*$ from $[A|B]$. The classical electrostatic model of the solvent leads to expressions for the energy contributions needed for achieving the interconversion of $[A|B]^*$ and $[A^+|B^-]^*$. There is nothing specific in this model concerning the preparation of the intramolecular degrees of freedom; nothing is said about how the nuclear displacements of the two molecules take place.

16.2   THE ELECTROSTATIC MODEL

We will start out by outlining the principles of the classical electrostatic model, and by using this model we will obtain the energy of the transition state (see Appendix F: Electrostatic Energy of a Polarized Dielectric). This determination of the energy of the transition state will allow us to obtain an activation energy and from transition state theory we will obtain a rate constant. The physical state of the system is completely defined by the charges of the molecules in $[A|B]^*$, (e_1^*, e_2^*), and $[A^+|B^-]^*$, (e_1, e_2), the radii of the molecules (a_1, a_2), and the polarization of the solvent due to the molecular charges. The solvent polarization $\mathbf{P}$ is the magnitude of the macroscopic dipole moment per unit volume in the solvent. The dipole moment is induced by the total electric field present, the field due to the charge distribution of the reacting system. The polarization vector consists of two components

$$\mathbf{P} = \mathbf{P}_{op} + \mathbf{P}_{in} \tag{16.5}$$

where $\mathbf{P}_{op}$ and $\mathbf{P}_{in}$ are the optical and inertial polarization vectors, respectively. The central feature of the dual transition state is based on the assumption that within $[A|B]^*$ and $[A^+|B^-]^*$ the ET is much faster than the time required for the solvent molecules to reorient themselves into the new charge state. Fluctuations in $\mathbf{P}_{in}$ bring the $[A|B]$ precursor into the transition state region where the dual electronic states exist. In this region the inertial polarization is not in equilibrium with the molecular charge distribution of $[A|B]^*$ or $[A^+|B^-]^*$. The optical polarization is permitted to change during ET and is always in equilibrium with the molecular charge distribution. The definition of the transition state is defined sufficiently by the value of the inertial polarization at the transition state. Here the energies of $[A|B]^*$ and $[A^+|B^-]^*$ must be the same. In addition the free energy of the transition state is actually an average over several solvent states weighted by appropriate Boltzmann factors. Accordingly these weighting factors decrease exponentially from that of the state of minimum free energy (the most probable state). As an approximation to the average free energy of the transition state, one uses the state of minimum free energy. The problem at hand is how to obtain an inertial polarization for which the two states are of the same energy and then to minimize the free energy for the transition state. This constraint reduces the free energy difference between the two states $[A|B]^*$ and $[A^+|B^-]^*$ to one connected to an entropy difference $-T\Delta S_e$. The entropy difference is due to differences in electronic degeneracies of $[A|B]^*$ and $[A^+|B^-]^*$; in most cases this is zero, but if it is not zero, it is always small. The contributions of the differences of the vibrational and rotational partition functions to the entropy difference are also taken to be zero on the basis of common atomic positions in the two states $[A|B]^*$ and $[A^+|B^-]^*$. We have, of course, assumed that the energies do not change with the changes in the spatial distribution of the transferred electron. Next we will consider the free energy of the transition state. In order to define the free energy of the states $[A|B]^*$ and $[A^+|B^-]^*$ we have to account for the energy needed to create a dipole moment in each volume element of the solvent and for the energies of the interaction of the ion and the solvent components of the electric field. The total electric fields for the two situations ($[A|B]^*$ and $[A^+|B^-]^*$) are

$$\mathbf{E} = \mathbf{E}_c + \mathbf{E}_{op} + \mathbf{E}_{in} \tag{16.6}$$

and

$$\mathbf{E}' = \mathbf{E}'_c + \mathbf{E}'_{op} + \mathbf{E}_{in} \tag{16.7}$$

The electric fields in the previous two equations stem from: (i) the electric field from the ions in the states $[A|B]^*$ ($\mathbf{E}_c$) and $[A^+|B^-]^*$ ($\mathbf{E}'_c$); (ii) the electric fields $\mathbf{E}_{op}$ and $\mathbf{E}'_{op}$ due to the optical polarization arising from the two charge distributions corresponding to the states $[A|B]^*$ and $[A^+|B^-]^*$, respectively; and (iii) the electric field ($\mathbf{E}_{in}$) due to the inertial polarization arising from the charge distribution of the state $[A|B]^*$. The local organization of the solvent by the electric field of the ions induces in the solvent a dipole moment per unit volume: $-\mu = \mathbf{P}\,dV$. The energies of each volume element in the presence of the total electric field must be a summation over all volume elements. The free energy stored in an induced dipole is given by

$$
\begin{aligned}
dF_1 &= \frac{\mu^2}{2\alpha} \\
&= \frac{\mathbf{P}^2_{op}}{2\alpha_{op}}\,dV + \frac{\mathbf{P}^2_{in}}{2\alpha_{in}}\,dV
\end{aligned}
\tag{16.8}
$$

where the polarizability per volume element $\alpha = \alpha_{op}\,dV$ or $\alpha = \alpha_{in}\,dV$ respectively.

The self-energy of the solvent dipole electric field interactions is

$$
\begin{aligned}
dF_2 &= -\frac{\mu}{2}\,(\mathbf{E}_{op} + \mathbf{E}_{in}) \\
&= -\frac{1}{2}\,\mathbf{P}(\mathbf{E}_{op} + \mathbf{E}_{in})\,dV
\end{aligned}
\tag{16.9}
$$

The energy of the solvent dipole-ion interactions is given by:

$$
\begin{aligned}
dF_3 &= -\mu\mathbf{E}_c \\
&= -\mathbf{P}\mathbf{E}_c\,dV
\end{aligned}
\tag{16.10}
$$

The self-energy of the two ions creating the electric field $\mathbf{E}_c$ is

$$dF_4 = \frac{\mathbf{E}_c^2}{8\pi} \, dV \tag{16.11}$$

Integration of these energy contributions over all volume elements gives the following expression for the free energy F:

$$\begin{aligned} F &= F_1 + F_2 + F_3 + F_4 \\ &= \frac{1}{2} \int \left[\frac{\mathbf{P}_{op}^2}{\alpha_{op}} + \frac{\mathbf{P}_{in}^2}{\alpha_{in}} - \mathbf{P}(\mathbf{E}_{op} + \mathbf{E}_{in}) - 2\mathbf{PE}_c + \frac{\mathbf{E}_c^2}{4\pi} \right] dV \end{aligned} \tag{16.12}$$

16.3   CHEMICAL ACTIVATION

Since the electronic degrees of freedom adjust themselves on a short timescale we assume an equilibrium situation for $\mathbf{P}_{op}$, i.e., $\mathbf{P}_{op} = \alpha_{op}\mathbf{E}$. Upon transfer of the electron, the nuclei and solvent configurations remain fixed in their respective positions and thus are in a nonequilibrium state. A nonequilibrium state exists for $\mathbf{P}_{in}$, and we assume that $\mathbf{P}_{in}$ is the same for the situation before and after the electron transfer. We can rewrite Eq. (16.12) as (see also Appendix F)

$$F = \frac{1}{2} \int \left[\frac{\mathbf{E}_c^2}{4\pi} - \mathbf{PE}_c + \mathbf{P}_{in} \left(\frac{\mathbf{P}_{in}}{\alpha_{in}} - \mathbf{E} \right) \right] dV \tag{16.13}$$

where $\mathbf{E}$ is the total field strength:

$$\mathbf{E} = \mathbf{E}_c + \mathbf{E}_{op} + \mathbf{E}_{in} \tag{16.14}$$

For the $[A|B]^*$ state we have

$$F([A|B]^*) = \frac{1}{2} \int \left[\frac{\mathbf{E}_c^2}{4\pi} - \mathbf{PE}_c + \mathbf{P}_{in} \left(\frac{\mathbf{P}_{in}}{\alpha_{in}} - \mathbf{E} \right) \right] dV \tag{16.15}$$

and for the $[A^+|B^-]^*$ state we have

$$F([A^+|B^-]^*) = \frac{1}{2} \int \left[\frac{\mathbf{E}_c'^2}{4\pi} - \mathbf{PE}_c' + \mathbf{P}_{in} \left(\frac{\mathbf{P}_{in}}{\alpha_{in}} - \mathbf{E}' \right) \right] dV \tag{16.16}$$

These free energy functionals are to be used in conjunction with the above-mentioned constraints: (i) $F([A|B]^*) - F([A^+|B^-]^*)$; (ii) $F(\mathbf{P}_{in})$ minimized with respect to variation in $\mathbf{P}_{in}$. Minimization of an integral subject to constraints is a problem dealt with by variational calculations. The solution is to make use of Lagrange multipliers m of the constraint combined with the function to be minimized:

$$\delta F^*(\delta \mathbf{P}_{in}) + m[\delta F^*(\delta \mathbf{P}_{in}) - \delta F(\delta \mathbf{P}_{in})] = 0 \qquad (16.17)$$

where δF^* is the difference of the free energy functional and where δ implies a variation. Here it should be remembered that $\mathbf{E}$ is not independent of $\mathbf{P}_{in}$. Following Marcus [1] we get:

$$\mathbf{P}_{in} = \alpha_{in}[\mathbf{E}' + m(\mathbf{E}' - \mathbf{E})] = \alpha_{in}[\mathbf{E} + (m + 1)(\mathbf{E}' - \mathbf{E})] \qquad (16.18)$$

Marcus succeeded in relating the field strength $\mathbf{E}$ to that of the charge distribution in vacuum $\mathbf{E}_c$ and found:

$$\mathbf{E}' - \mathbf{E} = \frac{1}{\epsilon_{op}} (\mathbf{E}'_c - \mathbf{E}_c) \qquad (16.19)$$

$$\mathbf{E}' = \frac{\mathbf{E}'_c}{\epsilon_{st}} - m(\mathbf{E}'_c - \mathbf{E}_c)\left(\frac{1}{\epsilon_{op}} - \frac{1}{\epsilon_{st}} \right) \qquad (16.20)$$

Using these relations in the expressions for the free energy we get:

$$F([A^+|B^-]^*) = \frac{1}{8\pi} \int dV \left(\frac{\mathbf{E}'^2_c}{\epsilon_{st}} + m^2(\mathbf{E}'_c - \mathbf{E}_c)^2 \left(\frac{1}{\epsilon_{op}} - \frac{1}{\epsilon_{st}} \right) \right) \qquad (16.21)$$

where ϵ_{op} and ϵ_{st} are the optical and static dielectric constants, respectively. They are related to the polarizabilities as

$$4\pi\alpha_{in} = \epsilon_{st} - \epsilon_{op} \qquad (16.22)$$

A similar expression is obtained for $F = F([A^+|B^-]^*)$ if $\mathbf{E}'_c$ is replaced by $\mathbf{E}_c$ and m by $m+1$. We are then able to determine m from the free energy difference between F^* and F, which is given by

$$F - F^* = \frac{1}{8\pi} \int dV \left(\frac{\mathbf{E}'^2_c - \mathbf{E}^2_c}{\epsilon_{st}} - (2m + 1)(\mathbf{E}'_c - \mathbf{E}_c)^2 \left(\frac{1}{\epsilon_{op}} - \frac{1}{\epsilon_{st}} \right) \right)$$

$$= \Delta F_0 + T\Delta S_e. \qquad (16.23)$$

where ΔF_0 is the standard free energy change and ΔS_e the entropy change connected to the electronic degrees of freedom. This equation can be used to determine the Lagrange multiplier m, i.e.,

$$-(2m + 1)\lambda = \Delta F_0 + T\Delta S_e + \frac{1}{8\pi} \int dV \frac{(\mathbf{E}_c^2 - \mathbf{E}_c'^2)}{\epsilon_{st}} \qquad (16.24)$$

where we have introduced the reorganization energy λ

$$\lambda = \frac{1}{8\pi} \int dV (\mathbf{E}_c' - \mathbf{E}_c)^2 \left(\frac{1}{\epsilon_{op}} - \frac{1}{\epsilon_{st}} \right) \qquad (16.25)$$

We are now able to find the free energy of activation:

$$E_A = F^* - W_{iso} = F^* - \left(\frac{e_1^{*2}}{a_1} + \frac{e_2^{*2}}{a_2} \right) \frac{1}{2\epsilon_{st}} \qquad (16.26)$$

where W_{iso} is the energy of the separated reactants and is calculated as the work required in charging the isolated ions while immersed in the solvent. The radii of the two ions are a_1 and a_2, respectively. Substituting the energy of F^* from above we obtain the activation energy (see Fig. 16.1) as:

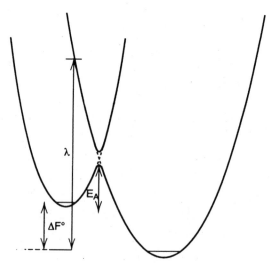

Fig. 16.1. Energy and reaction scheme for an electron transfer reaction in solution. The activation energy is E_A, ΔF^0 is the free energy of the reaction, and λ is the reorganization energy (the energy needed to go from the right configuration near its equilibrium position to the left configuration, or vice versa).

$$E_A = m^2\lambda + \frac{e_1^* e_2^*}{R\epsilon_{st}} \tag{16.27}$$

where the reorganization energy is

$$\lambda = (\Delta e)^2 \left(\frac{1}{2a_1} + \frac{1}{2a_2} - \frac{1}{R} \right) \left(\frac{1}{\epsilon_{op}} - \frac{1}{\epsilon_{st}} \right) \tag{16.28}$$

R is the sum of the two molecular radii and Δe is the charge transferred. We obtain an expression for m from Eq. (16.24)

$$-(2m + 1)\lambda = \Delta F^o + \frac{e_1 e_2 - e_1^* e_2^*}{R\epsilon_{st}} + T\Delta S_e \tag{16.29}$$

where $\Delta F^o = (W_{iso} - W_{iso}^*)$ is the free energy of the reaction. If the value of m is then substituted into the expression for the activation energy and the number of electrons transferred is set to one, we obtain (with $e_1 e_2 = e_1^* e_2^*$ and $T\Delta S_e \sim 0$)

$$E_A = \frac{(\Delta F^o + \lambda)^2}{4\lambda} + \frac{e^2 Z_1^* Z_2^*}{R\epsilon_{st}} \tag{16.30}$$

where Z_1^* and Z_2^* are the charges on the molecular ions and e is the electronic charge. We are now able to determine the rate constant according to transition state theory

$$k_{ET} = \frac{kT}{h} \exp\left(-\frac{E_{act}}{kT} \right) = \frac{kT}{h} \exp\left[-\frac{(\Delta F^o + \lambda)^2}{4\lambda kT} \right] \exp\left[-\frac{e^2 Z_1^* Z_2^*}{R\epsilon_{st} kT} \right] \tag{16.31}$$

which is the final result of Marcus theory. In summary the basic assumptions are:

(1) The states $[A|B]^*$ and $[A^+|B^-]^*$ are isoenergetic.
(2) The states $[A|B]^*$ and $[A^+|B^-]^*$ are defined in terms of isolated molecular properties.
(3) ET within the transition state region is not rate controlling.
(4) Dissociation of the transition state configuration is much slower than ET.
(5) The transition state configuration is not reached by internal activation of the molecules, it is only reached by rearrangements of solvent molecules.

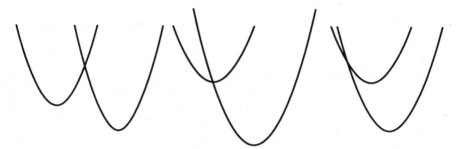

Fig. 16.2. Energy profiles for the three cases of an electron transfer reaction in solution. From left to right: $-\lambda < \Delta F^o$, $-\lambda = \Delta F^o$, and $-\lambda > \Delta F^o$.

(6) Although there exist a range of solvent orientations that permit electron transfer, the Boltzmann distribution over these states is ultimately ignored and only the lowest energy state for the transition state is considered.

(7) The system can be described in terms of a contact radius.

In order to understand the implications of the Marcus theory of electron transfer reactions we will consider three cases (see Fig. 16.2): (i) $-\lambda < \Delta F^o \leq 0$; (ii) $-\lambda = \Delta F^o$; (iii) $-\lambda > \Delta F^o$. The first case is termed the normal free energy region; note that the activation energy decreases as ΔF^o becomes more negative. In case (ii) we obtain an activationless rate and the observed rate constant is equal to the rate constant of diffusion. The third case is termed the inverted or abnormal region and the activation energy increases as the free energy of the reaction becomes more negative. The expression of the ET rate constant in the original presentation [1] did not include the influence of the electronic interactions at the transition state configuration. This aspect is considered in references [2–4].

REFERENCES

[1] R. A. Marcus, J. Chem. Phys. **24**,966,979(1956); **43**,679(1965).

[2] R. A. Marcus and N. Sutin, Biochim. Biophys. Acta **811**,265(1985).

[3] M. D. Newton and N. Sutin, Ann. Rev. Phys. Chem. **35**,437(1984).

[4] K. V. Mikkelsen and M. A. Ratner, Chem. Rev. **87**,113(1987).

EXERCISES

1. Derive Eq. (16.4).

2. Use the expressions $U_1 = U_1^0 + a_1(R - R_1^0)^2$ and $U_2 = U_2^0 + a_2(R - R_2^0)^2$

and show that $E_A = (\Delta F^0 + \lambda)^2/4\lambda$ yields $\lambda = U_2(R_1^0) - U_2^0$. What is the assumption one must make concerning the paramters a_1 and a_2? Explain why this assumption is reasonable.

3. Show in Fig. 16.2 the terms ΔF_0 and λ.

4. The rate constant for ET between a donor and an acceptor molecule can be written as:

$$k_{ET} = Z \exp(-E_A/kT) \qquad (16.32)$$

where $Z = 6.0 \cdot 10^{12}$ sec^{-1}, $E_A = z_1 z_2 e^2/\epsilon_{st}R + (1 + \Delta F/\lambda)^2(\lambda/4)$, and $\Delta F = \Delta F^0 + (|z_1| - |z_2| - 1)e^2/(\epsilon_{st}R)$. The static dielectric constant for the solvent is ϵ_{st}; and z_1 and z_2 are the charges on the donor and acceptor molecules, respectively. The distance between the donor and acceptor is R. The free energy of the reaction is given by ΔF^0, and λ is the reorganization energy.

a. Find an expression for k_{ET} for the reaction

$$NO_2^+ + NO_2 \rightarrow NO_2 + NO_2^+ \qquad (16.33)$$

For the solvent we have $\epsilon_{op} = 2$ and $\epsilon_{st} = 35$. Calculate k_{ET} as a function of the distance of R, i.e., for $a_i = 2$ Å, 3 Å, and 4 Å.

b. Consider the reactions

$$A^- + A \rightarrow A + A^- \qquad (16.34)$$
$$D^- + D \rightarrow D + D^- \qquad (16.35)$$

Find an expression for the reorganization energy for

$$D^- + A \rightarrow D + A^- \qquad (16.36)$$

if $\lambda_A = \lambda(A^-/A)$ and $\lambda_D = \lambda(D^-/D)$ are known.

5. The reorganization energy for an ET reaction is a sum of the inner-shell and the outer-shell reorganization energies. This exercise considers the inner-shell barrier for a pair of metal-ligand compounds ML_6^{2+} and ML_6^{3+}.

As seen in this chapter, the outer-shell reorganization energy is the energy required to change the configuration of the solvent in order to reach the transition state region. In this exercise the inner-shell reorganization energy is the energy required to change the metal-ligand distances from their equilibrium values to transition state values and we will make use of the harmonic approximation for the potential. The metal-ligand bond distances and force constants are d_2^o and f_2 for (ML_6^{2+}), and d_3^o and f_3 for (ML_6^{3+}).

a. Calculate the potential energy needed to change the metal-ligand bond distances in ML_6^{2+} and ML_6^{3+} from their equilibrium values to their transition state values, and simplify the expression by making use of energy conservation.

b. Obtain the inner-shell reorganization energy by minimizing the potential energy.

c. Generalize the expression from (b) to one for a complex that contains n identical ligands.

6. We consider a reaction between two ions: M and N. The two ions have an effective diffusion coefficient D and radii a_M and a_N, respectively.

a. Assume that the reaction takes place at the contact distance and determine an expression for the diffusion-controlled rate constant.

b. In the "steady-state" limit the total rate constant was measured to be k. Determine the rate constant for the chemical activated process between M and N using the results from a.

c. The two ions have the charges Z_M and Z_N. For a solvent with the optical dielectric constant ϵ_{op} and the static dielectric constant ϵ_{st} determine the dielectric solvation energy of each of the ions solvated in the solvent.

d. The chemical activated process is an ET process. Find an expression for the solvent reorganization energy and the activation energy of the reaction that involves the transfer of one electron. Explain the differences between the solvent reorganization energy for the ET reaction and the dielectric solvation energies of the two ions.

e. The free energy difference for the ET reaction is 0.0 kcal/mol. Determine, using the previous information, how the ET rate constant depends on the distance between the two ions. Omit the assumption of contact distance for the distance between N and M. Discuss whether the distance dependence is reasonable and which factors have been neglected.

APPENDIX A

UNITS

Various Constants in Molecular Units

$$\text{Mass unit} = 1\,\text{amu}$$

$$\text{Energy unit} = 1\hat{\epsilon} = 100\,\text{kJ/mol}$$

$$\text{Unit of length} = 1\,\text{Å}$$

$$\text{Unit of time} = 1\tau = 10^{-14}\,\text{sec}$$

$$\hbar = 0.06350781278\,\hat{\epsilon}\tau$$

$$\hbar^{-1} = 15.74609416\,\tau\,\text{Å}^{-2}\,\text{amu}^{-1}$$

$$1\,\text{eV} = 0.9648455078\,\hat{\epsilon}$$

$$k = 0.00008314410775\,\hat{\epsilon}/\text{K}$$

$$1\,\text{hartree} = 26.25499575\,\hat{\epsilon}$$

$$1\,\text{cm}^{-1} = 0.0001196265728\,\hat{\epsilon}/\hbar$$

$$e^2 = 13.89354158\,\hat{\epsilon}\,\text{Å}$$

$$1\,a_0 = 0.5291770644\,\text{Å}$$

$$1\,\text{Debye} = 10^{-18}\,\text{esu cm} = 0.7760183075\,\hat{\epsilon}^{1/2}\,\text{Å}^{3/2}$$

$$1\,\text{E} = 10^{-26}\,\text{esu cm}^2 = 0.7760183075\,\hat{\epsilon}^{1/2}\,\text{Å}^{5/2}$$

$$1\,\text{millidyne} = 1\,\text{md} = 6.022044\,\hat{\epsilon}/\text{Å}$$

$$\epsilon_0 = 8.854187825 \cdot 10^{-12}\,\text{F/m}$$

$$c \text{ (speed of light)} = 2.997924581 \cdot 10^8 \text{ m/sec}$$

$$1 \text{ e (charge of electron)} = 1.602189246 \cdot 10^{-19} \text{ C} = 4.803242524 \cdot 10^{-10} \text{ esu}$$

$$F \text{ (Faraday constant)} = 96484.5627 \text{ C/mol}$$

$$\hbar = 1.054588757 \cdot 10^{-34} \text{ J sec}$$

$$N_A = 6.02204531 \cdot 10^{23} \text{ mol}^{-1}$$

The constants and conversion factors are based upon the 1979 IUPAC values, "Pure and Applied Chemistry" **51**,1–41,(1979).

TABLE A.1. Energy Conversion Factors

	$\tilde{\nu}$ cm^{-1}	Energy E eV	Molar Energy E_m kJ/mol	Molar Energy E_m kcal/mol	Temperature K
$\tilde{\nu}$ (cm^{-1})	1	1.239842(−4)	1.196266(−2)	2.85914(−3)	1.43879
E (eV)	8065.54	1	96.4853	23.0605	1.16045(4)
E_m (kJ/mol)	83.5935	1.036427(−2)	1	0.239006	120.272
E_m (kcal/mol)	349.755	4.336411(−2)	4.184	1	503.217
T (K)	0.695039	8.61738(−5)	8.31451(−3)	1.98722(−3)	1

$E = h\nu = hc\tilde{\nu} = kT.$
1.239842(−4) means $1.239842 \cdot 10^{-4}$ and so on.

APPENDIX B

INTEGRALS AND FUNCTIONS

$$\int \frac{dx}{\sqrt{a - bx - cx^2}} = -\frac{1}{\sqrt{c}} \arcsin\left(-(b + 2cx)/\sqrt{b^2 + 4ac}\right) \qquad \text{(B.1)}$$

$$\int dx \sqrt{a + cx^2} = \frac{x}{2}\sqrt{a + cx^2} + \frac{a}{2\sqrt{-c}} \arcsin x\sqrt{-c/a}, \text{ where } c < 0 \quad \text{(B.2)}$$

$$\int_0^\infty x^{2n} \exp(-px^2)\, dx = \frac{(2n - 1)!!}{2(2p)^n} \sqrt{\frac{\pi}{p}} \qquad \text{(B.3)}$$

$$\int_0^\infty x^{2n+1} \exp(-px^2)\, dx = \frac{n!}{2p^{n+1}} \qquad \text{(B.4)}$$

$$\int_0^\infty \exp(-t)t^{z-1}\, dt = \Gamma(z) \qquad \text{(B.5)}$$

$$\frac{2}{\sqrt{\pi}} \int_0^x \exp(-t^2)\, dt = \operatorname{erf}(x) \qquad \text{(B.6)}$$

$$\operatorname{erfc}(x) = 1 - \operatorname{erf}(x) \qquad \text{(B.7)}$$

$$\mathrm{erfc}\,(x) \sim \frac{1}{\pi}\,\exp(-x^2)\left(\frac{\sqrt{\pi}}{x} + O(x^{-3})\right) \tag{B.8}$$

$$\int_{x_1 \geq 0} \int_{x_2 \geq 0} \cdots \int_{x_n \geq 0} dx_1 \ldots dx_n = \frac{h^n}{n!}\,, \text{ where } x_1 + x_2 + \ldots x_n \leq h \tag{B.9}$$

$$\int_{x_1 \geq 0} \int_{x_2 \geq 0} \cdots \int_{x_n \geq 0} dx_1 \ldots dx_n = \frac{\sqrt{\pi^n}}{\Gamma(n/2 + 1)}\,R^n,$$
$$\text{where } x_1^2 + x_2^2 + \ldots x_n^2 \leq R^2 \tag{B.10}$$

$$\int_{x_1 \geq 0} \int_{x_2 \geq 0} \cdots \int_{x_n \geq 0} dx_1 \ldots dx_n\, x_1^{d_1-1} \ldots x_n^{d_n-1}$$
$$= \frac{\displaystyle\prod_{i=1}^{n} \frac{q_i^{d_i}}{\alpha_i}\,\Gamma\!\left(\frac{d_i}{\alpha_i}\right)}{\Gamma\!\left(1 + \displaystyle\sum_{i=1}^{n} \frac{d_i}{\alpha_i}\right)}, \text{ where } \sum_{i=1}^{n} \left(\frac{x_i}{q_i}\right)^{\alpha_i} \leq 1 \tag{B.11}$$

$$\Gamma(\alpha) = \int_0^\infty \exp(-x)\,x^{\alpha-1}\,dx$$

$$\Gamma(x + 1) = x\Gamma(x)$$

$$\Gamma(\tfrac{1}{2}) = \sqrt{\pi} \tag{B.12}$$

$$(a + b)^n = \sum_{i=0}^{n} \binom{n}{i} a^i b^{n-i} \tag{B.13}$$

$$(1 - x)^{-q} = \sum_{k=0}^{\infty} \frac{(q + k - 1)!}{k!(q - 1)!}\,x^k \tag{B.14}$$

$$n! \sim \sqrt{2\pi n}\,n^n \exp(-n) \qquad \text{(Stirling's formula)} \tag{B.15}$$

APPENDIX C

LAPLACE TRANSFORM

The Laplace transform method is essentially a technique for transforming a partial differential equation into a total differential equation whose solution is a function of one independent variable. The total differential equation is solved and the solution is then transformed back to reintroduce the second independent variable.

If $F(t)$ is a known function for positive values of t, then the Laplace transform $L\{F(t)\}$ of $F(t)$ is defined as:

$$L\{F(t)\} = \int_0^\infty \exp(-st)F(t)\,dt = f(s) \qquad (C.1)$$

The existence of the transform is subject to the following conditions:

(1) $F(t)$ must be bounded for t in the interval $[0; \infty]$.
(2) $F(t)$ has a finite number of discontinuities
(3) $F(t)$ is of exponential order, so that the integrand is bounded for t approaching ∞.

The linear law holds for the Laplace transform:

$$L\{F + G\} = L\{F\} + L\{G\} = f(s) + g(s) \qquad (C.2)$$

Furthermore the Laplace transform for derivatives is

$$L\{dF(t)/dt\} = sf(s) - F(0) \tag{C.3}$$

which is easily seen from

$$L\{dF(t)/dt\} = \int_0^\infty \exp(-st)\,(dF(t)/dt)\,dt$$

$$= \exp(-st)F(t)|_0^\infty + s_0 \int_0^\infty \exp(-st)F(t)\,dt$$

$$= -F(0) + sf(s) \tag{C.4}$$

Generally one has for the nth derivative of $F(t)$

$$L\{F^{(n)}\} = s^n f(s) - s^{n-1}F(0) - s^{n-2}F^{(1)}(0) - \ldots - F^{(n-1)}(0) \tag{C.5}$$

The transformation is simple for differential operators not involving t

$$L\left\{ \frac{\partial F(x,t)}{\partial x} \right\} = \frac{\partial f(x,s)}{\partial x} \tag{C.6}$$

Variables other than t are considered to be constants during the conversion. The inversion of the Laplace transform normally makes use of tables containing Laplace transforms of common functions. If not found there, then the inversion is done by a convolution integral

$$L^{-1}\{f(s)g(s)\} = \int_0^\infty F(t-\tau)G(\tau)\,d\tau \tag{C.7}$$

Consider as an example the diffusion equation

$$\frac{\partial C(x,t)}{\partial t} = D\frac{\partial^2 C(x,t)}{\partial^2 x} \tag{C.8}$$

where there are an initial condition and boundary conditions; an initial condition ($t = 0$) and two boundary conditions for x, for example

$$C(x,0) = C^0 \qquad C(x,t) = C^0 \text{ for } x \text{ approaching } \infty \tag{C.9}$$

The third condition depends on the chemical system under consideration. We can find a partial solution using the two boundary conditions and the final evaluation depends on the third boundary condition. Laplace transform of the diffusion equations gives

$$sc(x, s) - C^0 = D \frac{\partial^2 c(x, s)}{\partial^2 x} \tag{C.10}$$

and

$$\frac{\partial^2 c(x, s)}{\partial x^2} - \frac{s}{D} c(x, s) = -\frac{C^0}{D} \tag{C.11}$$

The solution of this ordinary differential equation is

$$c(x, s) = \frac{C^0}{s} + a(s) \exp[-\sqrt{s/D}\, x] + b(s) \exp[\sqrt{s/D}\, x] \tag{C.12}$$

Due to the boundary condition for x, $b(s)$ must be zero. The third condition leads to an expression for $a(s)$ and the formal solution is

$$C(x, t) = C^0 + L^{-1}\{a(s) \exp(-\sqrt{s/D}\, x)\} \tag{C.13}$$

TABLE C.1. Laplace Transforms of Common Functions

$F(t)$	$f(s)$
A (constant)	A/s
$\exp(-at)$	$1/(s + a)$
$\sin(at)$	$a/(s^2 + a^2)$
$\cos(at)$	$s/(s^2 + a^2)$
$\sinh(at)$	$a/(s^2 - a^2)$
$\cosh(at)$	$s/(s^2 - a^2)$
t	$1/s^2$
$t^{n-1}/(n - 1)!$	$1/s^n$
$(\pi t)^{-1/2}$	$1/\sqrt{s}$
$2(t/\pi)^{1/2}$	$1/s^{3/2}$
$x/2(\pi k t^3)^{1/2} \exp(-x^2/4kt)$	$\exp(-\beta x)\ \ \beta = (s/k)^{1/2}$
$(k/\pi t)^{1/2} \exp(-x^2/4kt)$	$\exp(-\beta x)/\beta$
$\operatorname{erfc}[x/2(kt)^{1/2}]$	$\exp(-\beta s)/s$
$2(kt/\pi)^{1/2} \exp(-x^2/4kt) - x\operatorname{erfc}[x/2(kt)^{1/2}]$	$\exp(-\beta x)/s\beta$
$\exp(a^2 t)\operatorname{erfc}(at)^{1/2}$	$1/(s^{1/2}(s^{1/2} + a))$
$\operatorname{erfc}[k/2(t)^{1/2}]$	$\exp[-k(s)^{1/2}]/s$

APPENDIX D

STATISTICAL MECHANICS

A brief introduction provides the basic concepts of statistical mechanics: how the molecular behaviour or microscopic dynamics is related to the observed properties of a very large system. At present, it is impossible to solve the molecular equations of motion, either in terms of classical or quantum mechanics, for a many-body system containing moles of particles. Fortunately, the laws of thermodynamics have illustrated that large systems at thermodynamic equilibrium behave in a quite orderly fashion; one can characterize a macroscopic system with few variables, as illustrated by the ideal gas law. A simple-minded picture could be formulated as follows: The distinctive regularities of the macroscopic systems are bounded by the statistical laws governing the behavior of systems composed of a great number of microscopic particles. This leads to an obvious advantage: We do not have to evaluate the precise dynamics of the very large microscopic system but assume that probability statistics provides us with a description of the behavior of the macroscopic system.

As a start, imagine that it is possible to observe a many-body system in a particular microscopic state. The evolution of the system, governed either by classical or quantum mechanical equations of motion, will bring the system through different states belonging to a certain state space (phase space for classical mechanics, Hilbert space in the case of quantum mechanics). The evolution of the system will be dictated by a small number of variables, such as the total energy E, the total number of particles N, and the volume V. The constraints dictated by these variables control the microscopic states that the many-body system can attain through its trajectory in state space.

The observed quantity is given by

$$B_{\text{obs}} = \sum_{\nu} P_{\nu} B_{\nu} = \langle B \rangle \tag{D.1}$$

where

$$B_{\nu} = \langle \nu | B | \nu \rangle \tag{D.2}$$

is the expectation value for the quantity B when the system is in state ν and where P_{ν} is the probability of finding the many-body system in state ν.

The averaging operation (the weighted summation over B_{ν}) indicated by $\langle \rangle$ is called an ensemble average. An ensemble is a collection of all the possible microstates that are consistent with the conditions dictated by the characteristics of the macroscopic system. We will consider three different types of ensembles that are of relevance for the present textbook.

The microcanonical ensemble is an assembly of all states with fixed total energy E and fixed size, specified by the number of molecules N and the volume V. This is a closed and isolated system.

The canonical ensemble encompasses all states with a fixed size but without energy constraints, the latter gives rise to energy fluctuations. This is a closed system in contact with a heat bath.

The grand canonical ensemble represents the case where all the states can fluctuate in energy and size. This is an open system in contact with a heat bath.

Initially we will consider the microcanonical ensemble: a closed and isolated system with fixed total energy E and fixed size (V and N), and characterized by an equal likelihood of all the microscopic states at thermodynamic equilibrium. This gives a uniform distribution of microscopic states having the same energy and size. Letting $\Pi(N, V, E)$ represent the number of microstates with the size N and V and the energy between E and $E+dE$, allows us to write the probability of a macroscopic state ν for a system in equilibrium

$$P_{\nu} = \frac{1}{\Pi(N, V, E)} \tag{D.3}$$

which holds for all the states in the ensemble. For states with different sizes or with energies outside the interval specified $[E; E + dE]$, we have $P_{\nu} = 0$. This type of ensemble is of course appropriate for a system with fixed energy, volume, and number of molecules.

The canonical ensemble is a collection of all microstates with fixed size, fixed N and V, but fluctuating energy. The system is in contact with a heat bath and kept at thermal equilibrium at the temperature T. Basic assumptions about the heat bath are that it is of such great size that the energy levels of the bath are continuous, that $d\Pi/dE$ is well defined and also that the energy of the bath E_B is overwhelmingly larger than the energy of the system E_{ν}.

The system under investigation, described by a canonical ensemble, is appropriately viewed as a subsystem of a total system, the total system being represented as a microcanonical ensemble. The total system is isolated with fixed energy (E) and size (N, V) and is composed of the subsystem and the bath. The subsystem can exchange energy with the heat bath and thereby have a fluctuating energy.

The energy of the investigated system fluctuates because this system is in contact with the heat bath, but the total energy of the total system is constant.

$$E = E_B + E_\nu = \text{ constant} \tag{D.4}$$

The probability of finding the system under investigation in a definite state ν is proportional to $\exp(-\beta E_\nu)$, and normalization gives a constant of proportionality, which leads to the canonical or Boltzmann distribution law

$$P_\nu = \frac{\exp(-\beta E_\nu)}{Q} \tag{D.5}$$

where Q, the canonical partition function, is defined by

$$Q = \sum_\nu \exp(-\beta E_\nu) \tag{D.6}$$

The grand canonical ensemble describes an open system of volume V for which both energy and particle number can fluctuate from state to state. The probability of finding the investigated system in a definite state ν with N_ν particles and energy E_ν is given by

$$P_\nu = \frac{\exp(-\beta E_\nu + \beta\mu N_\nu)}{A} \tag{D.7}$$

where

$$A = \sum_\nu \exp(-\beta E_\nu + \beta\mu N_\nu) \tag{D.8}$$

which is the grand canonical partition function, and μ is the chemical potential. The partition function for N identical polyatomic molecules, Z, is

$$Z = \frac{Q^N}{N!} \tag{D.9}$$

from which the chemical potential can be derived using $F = -kT \ln Z$ and

$$\mu = \left. \frac{\partial F}{\partial N} \right|_{V,T} \tag{D.10}$$

i.e.,

$$\mu = -kT \ln \frac{Q}{N} \tag{D.11}$$

The partition function for a single molecule Q is

$$Q = \sum_s \exp(-\beta E_s) \tag{D.12}$$

where the energy of state s can be approximated as

$$E_s = E_t + E_r + E_v + E_e \tag{D.13}$$

i.e., a sum of the translational, rotational, vibrational, and electronic energy. If this separation is possible we can write the partition function Q as

$$Q = Q_t Q_r Q_v Q_e \tag{D.14}$$

i.e., as a product of partition functions for each degree of freedom. The separation between electronic states is usually large enough to replace the summation over electronic states with just the first term, i.e.,

$$Q_e = g_e \exp(-\beta E_e) \tag{D.15}$$

where g_e is the degeneracy factor and $E_e = -D_e$ for the ground state, where D_e is the dissociation energy for the electronic ground state. If the vibrational partition function is calculated with zero at the ground vibrational state $E_{v=0}$, E_e should be $-D_e + E_{v=0}$.

APPENDIX E

NOTES ON THE SOLVENT MODEL

This appendix provides an outline of the derivation leading to the expression in Eq. (13.11). The following integral equation was obtained by using Eq. (13.6) for the polarization field and the relation $\mathbf{E} = -\nabla V$

$$4\pi\epsilon_o \mathbf{P}(\mathbf{r}) + \chi\nabla \int da' \, \frac{\mathbf{P}(\mathbf{r}')\cdot\mathbf{n}'}{|\mathbf{r} - \mathbf{r}'|} = -\chi\nabla \int d\mathbf{r}' \, \frac{\rho(\mathbf{r}')}{|\mathbf{r} - \mathbf{r}'|} \tag{E.1}$$

and making use of the fact that the medium is isotropic (i.e., b is the radius of the cavity):

$$div\,\mathbf{P}(\mathbf{r}) = 0 \qquad |\mathbf{r}| > b\,(|\mathbf{r}| > |\mathbf{r}'|) \tag{E.2}$$

The position vector $\mathbf{r}$ refers to a point in the dielectric medium, therefore $|\mathbf{r}| > b$ and $|\mathbf{r}| > |\mathbf{r}'|$. This allows us to expand $|\mathbf{r} - \mathbf{r}'|$ in terms of the spherical harmonic functions $Y_{l,m}(\theta,\phi)$

$$\frac{1}{|\mathbf{r} - \mathbf{r}'|} = \sum_{l,m} \frac{4\pi}{2l+1} \left(\frac{r'}{r}\right)^l \frac{1}{r} (-1)^m Y_{l,m}(\theta,\phi) Y_{l,-m}(\theta',\phi') \tag{E.3}$$

$$\mathbf{r} = (r,\theta,\phi) \tag{E.4}$$

$$\mathbf{r}' = (r',\theta',\phi') \tag{E.5}$$

We introduce the spherical harmonic polynomials defined as

158

$$S_l^m(\mathbf{r}) = \sqrt{\frac{4\pi}{2l+1}} \; r^l Y_{l,m}(\theta,\phi) \tag{E.6}$$

We also define the spherical multipole moments for the charge distributions ρ and σ_p

$$M_l^m = \int d\mathbf{r}' \, \rho(\mathbf{r}') \, S_l^m(\mathbf{r}') \tag{E.7}$$

$$MP_l^m = \int d\mathbf{a}' \, \sigma_p(\mathbf{r}') \, S_l^m(\mathbf{r}') = \int d\mathbf{a}' \, \mathbf{n}' \cdot \mathbf{P}(\mathbf{r}') \, S_l^m(\mathbf{r}') \tag{E.8}$$

Using the Eqs. (E.5), (E.6), (E.7), and (E.8) we are able to rewrite Eq. (E.1)

$$4\pi\epsilon_o \mathbf{P}(\mathbf{r}) + \chi \nabla \left\{ \sum_{l,m} \left(\frac{4\pi}{2l+1} \right)^{\frac{1}{2}} r^{-(l+1)} Y_{l,m}(\theta,\phi)(-1)^m MP_l^{-m} \right\}$$

$$= -\chi \nabla \left\{ \sum_{l,m} \left(\frac{4\pi}{2l+1} \right)^{\frac{1}{2}} r^{-(l+1)} Y_{l,m}(\theta,\phi)(-1)^m M_l^{-m} \right\} \tag{E.9}$$

The gradient operator is represented, in spherical coordinates, as

$$\nabla = -\mathbf{n} \frac{\partial}{\partial r} + \frac{i}{r} \mathbf{n} \times \mathbf{L} \tag{E.10}$$

$$\mathbf{n} = \frac{\mathbf{r}}{r} \tag{E.11}$$

$$\mathbf{L} = \frac{1}{i}(\mathbf{r} \times \nabla) \tag{E.12}$$

$$L^2 Y_{l,m}(\theta,\phi) = l(l+1)Y_{l,m}(\theta,\phi) \tag{E.13}$$

The gradient of $(r^{l+1}Y_{l,m})$ is given by

$$\nabla(r^{l+1}Y_{l,m}) = (l+1)r^{-(l+2)}Y_{l,m}(\theta,\phi)\,\mathbf{n} + ir^{-(l+2)}\mathbf{n} \times \mathbf{L}Y_{l,m}(\theta,\phi) \tag{E.14}$$

The next step involves finding the dot product between Eq. (E.9) and $\mathbf{n}$. We make use of the expression in Eq. (E.14) and obtain

$$4\pi\epsilon_o \mathbf{n} \cdot \mathbf{P(r)} + \chi \sum_{l,m} \left(\frac{4\pi}{2l+1}\right)^{\frac{1}{2}} (l+1)r^{-(l+2)}Y_{l,m}(\theta,\phi)(-1)^m MP_l^{-m}$$

$$= -\chi \sum_{l,m} \left(\frac{4\pi}{2l+1}\right)^{\frac{1}{2}} (l+1)r^{-(l+2)}Y_{l,m}(\theta,\phi)(-1)^m M_l^{-m} \qquad \text{(E.15)}$$

Finally we multiply both sides of Eq. (E.15) by $(S_l^m)^*$ and integrate with respect to the cavity surface

$$(MP_l^m)^* = (M_l^m)^* \frac{(-\chi/\epsilon_o)(l+1)}{(2l+1) + \chi\,\epsilon_o(l+1)} \qquad \text{(E.16)}$$

The polarization charge density on the cavity surface is given by

$$\sigma_p = \mathbf{n} \cdot \mathbf{P(r)}$$

$$= -\frac{\chi}{4\pi\epsilon_o} \sum_{l,m} \left(\frac{4\pi}{2l+1}\right)^{\frac{1}{2}}$$

$$\cdot \frac{(l+1)(2l+1)}{(\chi/\epsilon_o)(l+1) + (2l+1)} r^{-(l+2)}Y_{l,m}(\theta,\phi)(-1)^m M_l^{-m} \qquad \text{(E.17)}$$

We are now able to calculate the polarization energy given in Eq. (13.18).

APPENDIX F

ELECTROSTATIC ENERGY OF A POLARIZED DIELECTRIC

The polarization of a dielectric medium consists of electronic, atomic, and orientational contributions. The electronic component of the polarization, the optical polarization ($P_{op}(r)$), is due to distortions of the electronic degrees of freedom of the solvent molecules. The atomic and orientational components give rise to the inertial polarization ($P_{in}(r)$). The total polarization is a vector sum

$$P(r) = P_{in}(r) + P_{op}(r) \qquad (F.1)$$

An equilibrium state of the system is uniquely defined by the electric field vector ($E(r)$, determined by the electrostatic potential) and the electric displacement vector ($E_c(r)$, determined by the charge distribution)

The polarization is a function of the electric field vector $E(r)$ and is expressed as

$$P(r) = \alpha E(r) \qquad (F.2)$$

where α is the polarizability of the medium. For an equilibrium state the optical and inertial polarization vectors can be expressed as

$$P_{op}(r) = \alpha_{op} E(r) \qquad (F.3)$$

$$P_{in}(r) = \alpha_{in} E(r) \qquad (F.4)$$

where the polarizabilities are written as

161

$$\alpha_{op} = \frac{(\epsilon_{op} - 1)}{4\pi} \tag{F.5}$$

$$\alpha_{in} = \frac{(\epsilon_{st} - \epsilon_{op})}{4\pi} \tag{F.6}$$

where ϵ_{op} and ϵ_{st} are the optical and static dielectric constants of the medium, respectively. The potential $(V(\mathbf{r}'))$ at a point $\mathbf{r}'$ is given by

$$V(r') = \int \frac{\rho(\mathbf{r})}{|\mathbf{r} - \mathbf{r}'|} \, dV + \int \frac{\sigma(\mathbf{r})}{|\mathbf{r} - \mathbf{r}'|} \, dS + \int \mathbf{P}(\mathbf{r}) \cdot \nabla_r \frac{1}{|\mathbf{r} - \mathbf{r}'|} \, dV \tag{F.7}$$

where $\rho(\mathbf{r})$ is the charge density per unit volume dV at the point $\mathbf{r}$ of the system and $\sigma(\mathbf{r})$ is the charge density per unit area at a surface element dS of an interface of the system. The polarization vector $\mathbf{P}$ is the sum of the two polarization vectors. The electric displacement vector is defined as the electric field due a charge distribution in vacuum and given as

$$\mathbf{E}_c(\mathbf{r}') = -\nabla_r \left(\int \frac{\rho(\mathbf{r})}{|\mathbf{r} - \mathbf{r}'|} \, dV + \int \frac{\sigma(\mathbf{r})}{|\mathbf{r} - \mathbf{r}'|} \, dS \right) \tag{F.8}$$

or as

$$\text{div } \mathbf{E}_c = 4\pi\rho \tag{F.9}$$

The general expression for charging a dielectric medium in equilibrium is

$$W = \frac{1}{8\pi} \int \mathbf{E} \cdot \mathbf{E}_c \, dV = \frac{1}{8\pi\epsilon_{st}} \int \mathbf{E}_c^2 \, dV \tag{F.10}$$

This expression is obtained with the conditions

$$\frac{\delta W}{\delta P_{op}} = 0, \qquad \frac{\delta W}{\delta P_{in}} = 0 \tag{F.11}$$

A nonequilibrium state is specified by the polarization $\mathbf{P}$, the electric field vector $\mathbf{E}$, and the electric displacement vector $\mathbf{E}_c$. In this case the optical polarization vector is in equilibrium with the electric field, i.e., $\mathbf{P}_{op} = \alpha_{op}\mathbf{E}$ whereas the inertial polarization possesses an arbitrary value. The electrostatic charging energy is expressed as

$$W' = \frac{1}{8\pi} \int \mathbf{E}_c^2 \, dV - \frac{1}{2} \int \left\{ \mathbf{P}\mathbf{E}_c + \mathbf{P}_{in} \left(\frac{\mathbf{P}_{in}}{\alpha_{in}} - \mathbf{E} \right) \right\} \, dV \tag{F.12}$$

subject to the condition that

$$\frac{\delta W}{\delta P_{in}} = 0 \tag{F.13}$$

The polarization energy of the dielectric medium U is obtained by subtracting from the electrostatic charging energy (W and W') the charging energy of the system with the same charge distribution in vacuum (W_{vac}). The expression for W_{vac} is

$$W_{vac} = \frac{1}{8\pi} \int \mathbf{E}_c^2 \, dV \tag{F.14}$$

For an equilibrium state the polarization energy of the dielectric is given as Eqs. (F.10–F.14)

$$U = -\frac{1}{8\pi} \left(1 - \frac{1}{\epsilon_{st}} \right) \int \mathbf{E}_c^2 \, dV \tag{F.15}$$

and for a nonequilibrium state we have Eqs. (F.12–F.14)

$$U' = -\frac{1}{2} \int \left\{ \mathbf{P}\mathbf{E}_c + \mathbf{P}_{in} \left(\frac{\mathbf{P}_{in}}{\alpha_{in}} - \mathbf{E} \right) \right\} \, dV. \tag{F.16}$$

The latter equation can be transformed to

$$U' = -\frac{1}{8\pi} \left(1 - \frac{1}{\epsilon_{op}} \right) \int \mathbf{E}_c^2 \, dV$$
$$- \frac{1}{\epsilon_{op}} \int \mathbf{P}_{in} \mathbf{E}_c \, dV + \frac{2\pi\epsilon_{st}}{\epsilon_{op}(\epsilon_{st} - \epsilon_{op})} \int \mathbf{P}_{in}^2 \, dV \tag{F.17}$$

where in the transformation we make use of Eq. (F.6)

$$\mathbf{E} = \frac{\mathbf{E}_c - 4\pi \mathbf{P}_{in}}{\epsilon_{op}} \tag{F.18}$$

and

$$\mathbf{P}_{op} = \frac{1}{4\pi} \left(1 - \frac{1}{\epsilon_{op}} \right) \mathbf{E}_c - \left(1 - \frac{1}{\epsilon_{op}} \right) \mathbf{P}_{in}. \tag{F.19}$$

APPENDIX G

ANSWERS TO EXERCISES

Chapter 2

1. $D_e = 5.074$ eV and $\beta = (1/\hbar)\sqrt{2\mu x_e \hbar \omega_e} = 1.878$ Å for H_2

2. Use, for example, that the kinetic energy is given as: $\frac{1}{2}(\mu \dot{R}^2 + m\dot{r}^2)$, where $r = R_{BC}$ and $R = R_{AB} + (m_C/m_B + m_C)R_{BC}$; $\mu = m_A(m_B + m_C)/(m_A + m_B + m_C)$; and $m = m_B m_C/(m_B + m_C)$. The cross term vanishes if $R_{BC} = y/\sin(\xi)$ and $R_{AB} = \sqrt{\tilde{\mu}/\mu}(x - R_{BC}\cos\xi)$. The scaling parameter $c = \sqrt{m_A(m_B + m_C)/m_C(m_A + m_B)}$.

Chapter 3

1. From Fig. 3.3 we have $\psi = b/R$ (R large), i.e., $\dot{\psi} = -(\frac{b}{R^2})\dot{R} = bv/R^2$.

2. $L = \mu v R \sin \psi = \mu b v$.

Chapter 4

1. cm^3/sec.

2. It follows from the derivation that $n_A n_B k(T)$ is the rate at which A particles disappear.

3.
$$\sigma_r = \frac{n\pi}{2} C^{2/n} \left(\frac{n}{2} - 1\right)^{2/n-1} E_{kin}^{-2/n} \tag{G.1}$$

and the rate constant

$$k_r(T) = \sqrt{8/\pi\mu}\; \frac{n\pi}{2}\, C^{2/n} \left(\frac{n}{2} - 1\right)^{2/n-1} \Gamma\left(2 - \frac{2}{n}\right) (kT)^{1/2-2/n} \qquad (G.2)$$

For $n = 4$ the rate constant is independent of temperature and equal to 15.8 10^{-10} cm^3/sec ($k_r = \pi\sqrt{8C/\mu}$). The theoretically determined cross sections are 14.7 Å^2, 8.5 Å^2, and 26.9 Å^2; respectively; ($\sigma_r = \pi\sqrt{4C/E_{kin}}$, where $C = 0.5\alpha e^2 = 550$ kJÅ^4/mol).

4. 1 GK $= \sqrt{8kT/\pi\mu}\, \pi b_{max}^2$. At 300 K, 1 GK $= 5.0 \cdot 10^{-10}$ cm^3/sec.

$$k(T) = \sqrt{8kT/\pi\mu}\, \pi b_{max}^2 \exp(-\beta E_0)(1 + \beta E_0).$$

At 300 K, $k(T) = 2.1 \cdot 10^{-13}$ cm^3/sec; ($1.9 \cdot 10^{-14}$ cm^3/sec if the factor $(1 + \beta E_0)$ is omitted).

5. The factor of two arises as a consequence of the symmetry around R_0.

6. $d\sigma' = d^2/4$ and $d\sigma' = (a/4E)^2 \csc^4(\theta/2)$.

Chapter 5

1. $$\langle v_x \rangle = \frac{1}{m} \int_0^\infty dp_x\, p_x \sqrt{\frac{1}{2\pi mkT}} \exp(-p_x^2/2mkT) = \sqrt{\frac{kT}{2\pi m}} \qquad (G.3)$$

2. $$Q_{rot} = \sum_j (2j + 1)\exp[-j(j+1)\hbar^2/2IkT]$$

$$= \int du\, \exp(-u\hbar^2/2IkT) = \frac{2IkT}{\hbar^2} \qquad (G.4)$$

4. Use

$$dn_x = \frac{L}{2\pi}\, dk_x = \frac{L}{h}\, dp_x = \int_0^L \frac{dx\, dp_x}{h}$$

Chapter 6

1. $\sigma_{symm} = 2$.

Rate constants: 200 K, $2.88 \cdot 10^{-17}$; 300 K, $2.88 \cdot 10^{-15}$; 500 K, $1.2 \cdot 10^{-13}$; 1000 K, $7.14 \cdot 10^{-12}$ cm^3/sec.

Cross-sections: $1.87 \cdot 10^{-6}$; $1.53 \cdot 10^{-4}$; $4.93 \cdot 10^{-3}$; and 0.208 Å^2.

Probabilities: $0.78 \cdot 10^{-6}$; $0.64 \cdot 10^{-4}$; 0.0021; 0.087.

2. a. $\Delta E_0 = 12.5$ kcal/mol.

b. $\sigma_{symm} = 4$.

c. $k(T) = \sigma_{symm} \dfrac{kT}{\hbar} \exp(-\Delta E_0/kT) \dfrac{Q^{\#}_{vib}}{Q_{vib}Q_{trans}} \dfrac{Q^{\#}_{rot}}{Q_{rot}}$

$Q_{trans} = (2\mu\pi kT)^{3/2}/h^3$

$Q_{vib} = \Pi_i 1/(1 - \exp(-\hbar\omega_i/kT))$

d. Rate constants: 300 K, $k = 2.6 \cdot 10^{-20}$ cm^3/sec; 1000 K, $k = 1.6 \cdot 10^{-13}$ cm^3/sec.

e. For example: tunnel-effect, ΔE_0 too large.

f. Isotope-effects are involved in partition functions, zero point correction, and tunnel correction factor.

3. $\sigma_{symm} = 1$,

Rate constants: 300 K, $4.2 \cdot 10^{-15}$ cm^3/sec; 1000 K, $3.2 \cdot 10^{-12}$ cm^3/sec.

Chapter 7

1. $\kappa \sim 2.3$ at 200 K.

Chapter 8

1. Use partial integration.

2. The difference of a factor π is due to the "incorrect" integration over phase space. If Euler angles and their conjugate momenta are used then we have:

$$Q_{rot} = \frac{1}{h^3} \int_0^\pi d\theta \int_0^{2\pi} d\phi \int_0^{2\pi} d\psi \int dp_\theta \int dp_\phi \int dp_\psi \exp(-E_{rot}/kT) \qquad (G.5)$$

Introducing

$$E_{rot} = \frac{1}{2}\left(\frac{P_x^2}{I_a} + \frac{P_y^2}{I_b} + \frac{P_z^2}{I_c} \right)$$

and

$$dp_\theta \, dp_\phi \, dp_\psi = \sin\theta \, dP_x \, dP_y \, dP_z$$

we get the expression in Table 5.2.

3.
$$k_{\text{uni}}^0(T) = [M]Z^0 \int_{E_0}^{\infty} dE \, \frac{\rho(E)}{Q_{\text{vib}}} \exp(-E/kT) \qquad \text{(G.6)}$$

where

$$\rho(E) = \frac{E^{(s-1)}}{(s-1)! \Pi_{i=1}^s \hbar\omega_i}$$

Thus we get

$$k_{\text{uni}}^0(T) = [M]Z^0 \, \frac{1}{Q_{\text{vib}}(s-1)! \Pi_{i=1}^s \beta\hbar\omega_i} \, \Gamma(s, \beta E_0) \qquad \text{(G.7))}$$

where $\beta = 1/kT$. Using the asymptotic formula we have

$$k_{\text{uni}}^0(T) = [M]Z^0 \, \frac{(\beta E_0)^{s-1}}{(s-1)! Q_{\text{vib}} \Pi_{i=1}^s \beta\hbar\omega_i} \, \exp(-\beta E_0)$$

$$= \omega \, \frac{(\beta E_0)^{s-1}}{(s-1)!} \, \exp(-\beta E_0) \qquad \text{(G.8)}$$

At high temperatures

$$k_{\text{uni}}^\infty = \frac{1}{2\pi} \, \frac{\displaystyle\prod_{i=1}^{18} \omega_i}{\displaystyle\prod_{i=1}^{17} \omega_i^{\#}} \, \exp(-\beta E_0) = 10^{13.16} \exp(-\beta E_0) \, \text{sec}^{-1} \qquad \text{(G.9)}$$

A more correct expression is

$$k_{uni}^{\infty} = \frac{kT}{\hbar} \frac{\prod_{i=1}^{18}[1 - \exp(-\hbar\omega_i/kT)]}{\prod_{i=1}^{17}[1 - \exp(-\hbar\omega_i^{\#}/kT)]} \exp(-\beta E_0) \qquad (G.10)$$

and we define

$$E_{\infty} = -k \frac{d[\ln(k_{uni}^{\infty})]}{d(1/T)} = E_0 + kT + \sum_i \frac{\hbar\omega_i^{\#}}{\exp(\hbar\omega_i^{\#}/kT) - 1}$$

$$- \sum_i \frac{\hbar\omega_i}{\exp(\hbar\omega_i/kT) - 1} \qquad (G.11)$$

We get $E_{\infty} = 235.6$ kJ/mol. Furthermore

$$\ln A_{\infty} = \ln k_{uni}^{\infty} + \frac{E_{\infty}}{kT} \qquad (G.12)$$

giving $A_{\infty} = 10^{13.3}$ sec^{-1}.

5. We can rewrite $(i - m + s - 1)!i!/(i - m)!(i + s - 1)!$ as

$$\frac{(i - m + s - 1)(i - m + s - 2)\dots(i - m + 1)}{(i + s - 1)(i + s - 2)\dots(i + 1)}$$

where the numerator and denominator contain $s - 1$ terms. We now divide by i^{s-1}, i.e., each term by i, and get

$$\frac{\prod_{k=1}^{s-1}\left(1 - \frac{m + k - s}{i}\right)}{\prod_{k=1}^{s-1}\left(1 + \frac{s - k}{i}\right)} \sim \frac{\exp\left(-\sum_{k=1}^{s-1}\frac{m + k - s}{i}\right)}{\exp\left(\sum_{k=1}^{s-1}\frac{s - k}{i}\right)}$$

$$\sim \exp[-(s - 1)m/i] \sim \left(1 - \frac{m}{i}\right)^{s-1} \qquad (G.13)$$

where $m/i = E_0/E$, i.e.,

$$k_a(E) = A \left(1 - \frac{E_0}{E} \right)^{s-1} \tag{G.14}$$

Thus we get

$$k_{\text{uni}}(T) = \int_{E_0}^{\infty} dE \; \frac{A \left(1 - \dfrac{E_0}{E} \right)^{s-1} P(E,T)}{1 + A \left(1 - \dfrac{E_0}{E} \right)^{s-1} \bigg/ \omega} \tag{G.15}$$

where

$$P(E,T) = \frac{\rho(E)}{Q} \exp(-E/kT) \tag{G.16}$$

and

$$\rho(E) = \frac{E^{s-1}}{\Gamma(s) \displaystyle\prod_{i=1}^{s} \hbar\omega_i} \tag{G.17}$$

Thus in the high-pressure limit we have $\omega \to \infty$, i.e.,

$$k_{\text{uni}}^{\infty}(T) = \frac{A}{Q} \frac{(kT)^s}{\prod \hbar\omega_i} \exp(-\beta E_0) \sim A \exp(-\beta E_0) \tag{G.18}$$

where we have used

$$\Gamma(s) = \int_0^{\infty} dx \; x^{s-1} \exp(-x).$$

In the low-pressure limit we have

$$k_{\text{uni}}^0(T) = \frac{\omega}{Q\Gamma(s)} \frac{(kT)^s}{\prod \hbar\omega_i} \Gamma(s, x_0) \tag{G.19}$$

where $x_0 = E_0/kT$. For large values of x_0 we have $\Gamma \sim x_0^{s-1} \exp(-x_0)$, i.e.,

$$k_{uni}^0(T) = \frac{\omega}{Q\Gamma(s)} \frac{(kT)^s}{\prod \hbar\omega_i} \left(\frac{E_0}{kT}\right)^{s-1} \exp(-E_0/kT)$$

$$\sim \frac{\omega}{\Gamma(s)} \left(\frac{E_0}{kT}\right)^{s-1} \exp(-E_0/kT) \tag{G.20}$$

For $s = 10$ we have $k_{uni}^0 = 9.0 \cdot 10^{-4}$ sec^{-1} and with $s = 15$, $k_{uni} = 0.6$ sec^{-1}.

6. We introduce $k_1(E)/k_2(E) = P_j$ and obtain

$$k_{uni}(T) = A \sum_{i>m} P_i \frac{(i-m+s-1)!i!}{(i-m)!(i+s-1)!} \left/ \left[1 + A \frac{(i-m+s-1)!i!}{\omega(i-m)!(i+s-1)!}\right]\right. \tag{G.21}$$

Using $g_i = (i+s-1)!/i!(s-1)!$ we get

$$P_i = g_i \exp(-i\hbar\omega/kT)[1 - \exp(-\hbar\omega/kT)]^s \tag{G.22}$$

and hence

$$k_{uni}(T) = (1 - \exp[-\hbar\omega/kT])^s \times A \exp(-\beta E_0) \sum_{p=0}$$

$$\cdot \frac{[(p+s-1)!/p!(s-1)!]\exp(-p\hbar\omega/kT)}{1 + A[(p+m)!(p+s-1)!/\omega p!(p+m+s-1)!]} \tag{G.23}$$

where m is identified through the relation $E_0 = m\hbar\omega$.

7. $k(T) = 5.65 \ 10^{-5}$ sec^{-1} at $T = 1000$ K and $k(T) = 9.60 \ 10^4$ sec^{-1} at 2000 K.

Chapter 9

1. $\mathbf{r}' = (1-\epsilon_1)\mathbf{r} - \mathbf{R}$ and $\mathbf{R}' = -\epsilon_3\mathbf{R} - \mathbf{r}[1-\epsilon_3(1-\epsilon_1)]$ where $\epsilon_3 = m_A/(m_A+m_B)$.

2. $$n = -\frac{1}{2} + \frac{2}{h}\int_{r_-}^{r_+} dr\sqrt{2m(E - \tfrac{1}{2}kr^2)} = 2\pi E/h\omega$$

3. Introduce $y = \exp[-\beta(r - r_e)]$ and expand $1/r^2$ in a power series in y. Use that

$$\Delta t = \int dt = \text{const} \int dy \frac{1}{y\sqrt{a + by + cy^2}} \tag{G.24}$$

is uniform.

Chapter 10

1. Use perturbation theory with $\phi_1(t_0) = 1$ and $\phi_2(t_0) = 0$.

2.
$$P_{12} = \exp\left(-2\pi H_{12}^2 e^2 / \hbar v \Delta E^2\right) \tag{G.25}$$

Introducing the parameters we get $P_{12} = 0.84$.

3. a. $v_c = 2\pi H_{12}^2 R_c^2 / e^2 \hbar$.

    b. $R_c = 11.2$ Å, $v_c = 0.132$ Å$/10^{-14}$ sec.

    c. $P = 2P_{12}(1 - P_{12})$

    d. $v = 1.443 v_c$ or $P = 0.453$.

Chapter 11

1. Number of collisions per unit area: $\frac{1}{2}nv_z$. Transfer of momentum $2mv_z$.

2. Substitute $\alpha_i = \alpha_0 C x^i$ in $\sigma = \sigma_m \sum_i i\alpha_i$ and use $\sum_{i=1} ix^i = x/(1-x)^2$, $1 = \sum_{i=0} \alpha_i$, and $\sum_{i=1} x^i = x/(1-x)$.

3. We find $V_m = 0.31$ cm^3, $\sigma_m = 0.0763$ molecule/Å^2.

4. $D_0 = 2.08 \cdot 10^{-3}$ cm^2/sec. The model assumes that the motion in the x-direction is independent of the motion in the y-direction.

Chapter 12

1. Use

$$(a+b)^m = \sum_{r=0}^{m} \frac{m!}{r!(m-r)!} a^r b^{m-r}$$

2. Introducing $v = m/2$ and $k = v - r$ we have

$$P(v,r) = P_0 \frac{(v)!(v)!}{(v-k)!(v+k)!} = P_0 \frac{v(v-1)\ldots(v-k+1)}{(v+1)\ldots(v+k)} \tag{G.26}$$

where

$$P_0 = \frac{(2v)!}{v!v!} \left(\frac{1}{2}\right)^{2v}$$

Stirling's formula yields

$$P_0 \sim \frac{1}{\sqrt{\pi \nu}} \tag{G.27}$$

Divide numerator and demoninator by ν^k and use $1 + j/\nu \sim \exp(j/\nu)$. We then get

$$P(\nu, k) \sim \frac{\exp(-k^2/\nu)}{\sqrt{\pi \nu}} \tag{G.28}$$

Introducing $\nu = m/2 = t/2\tau$ and $k = m/2 - r = -x/2\lambda$ we have

$$P(x, t) = \frac{\lambda}{\sqrt{D\pi t}} \exp(-x^2/4Dt) \tag{G.29}$$

Using $\lambda = dx/2$ we get the desired result.

3. $D_{H^+} = 9.314 \cdot 10^{-5}$, $D_{Li^+} = 1.03 \cdot 10^{-5}$, and $D_{I^-} = 2.05 \cdot 10^{-5}$ cm^2/sec.

4. The solution can be obtained as

$$v(t) = \frac{1}{m} \int_{t_0}^{t} dt' \, F(t') \exp\left(\frac{\gamma}{m}(t' - t)\right) \tag{G.30}$$

from which

$$\langle v(t)v(t') \rangle = \frac{1}{m^2} \int_{-\infty}^{t} ds \, \exp\left[-\frac{\gamma}{m}(t - s)\right] \int_{-\infty}^{t'} ds'$$
$$\cdot \exp\left[-\frac{\gamma}{m}(t' - s')\right] \langle F(s)F(s') \rangle$$
$$= \frac{kT}{m} \exp\left(-\frac{\gamma}{m}|t' - t|\right) \tag{G.31}$$

Chapter 13

1. a.
$$E_{pol} = -\frac{1}{2a} q^2 \frac{\epsilon - 1}{\epsilon} \tag{G.32}$$

b.
$$E_{pol} = -\frac{1}{2a^3} d_{10}^2 \frac{2(\epsilon - 1)}{1 + 2\epsilon} \tag{G.33}$$

c.
$$E_{pol} = -\frac{1}{2a^5} Q_{20}^2 \frac{3(\epsilon - 1)}{2 + 3\epsilon} \tag{G.34}$$

2. The solvent-induced energy shift related to absorption is

$$\Delta E_{abs} = -\tfrac{1}{2}f_1(\epsilon_{op})(\mu_{ex}\mu_{ex} - \mu_{gs}\mu_{gs}) - \tfrac{1}{2}f_1(\epsilon_{st}, \epsilon_{op})\mu_{gs}(\mu_{ex} - \mu_{gs}) \qquad (G.35)$$

The solvent-induced energy shift related to fluorescence is

$$\Delta E_{flu} = -\tfrac{1}{2}f_1(\epsilon_{op})(\mu_{gs}\mu_{gs} - \mu_{ex}\mu_{ex}) - \tfrac{1}{2}f_1(\epsilon_{st}, \epsilon_{op})\mu_{ex}(\mu_{gs} - \mu_{ex}) \qquad (G.36)$$

which gives

$$\delta = \Delta E_{abs} + \Delta E_{flu} = \tfrac{1}{2}f_1(\epsilon_{st}, \epsilon_{op})(\mu_{ex} - \mu_{gs})^2 \qquad (G.37)$$

3. **a.** Plot the relative excitation energies as a function of $\tfrac{1}{2}f_1(\epsilon_{st}^b, \epsilon_{st}^a)$. The curves giving straight lines correspond to excitations where Eqs. (13.36) and (13.37) are valid.

   b. The slope of the line, $\alpha = \mu(\mu* - \mu)$ where μ and $\mu*$ are the dipole moments in the ground state and the excited state, respectively. This gives $\mu* = \alpha/\mu + \mu$.

   c. The solvent polarization energy for the excited state in a solvent with the static dielectric constant ϵ_{st} is given by

$$E_{ex} = -\frac{1}{2}f_1(\epsilon_{st})\left(\frac{\alpha}{\mu} + \mu\right)^2 \qquad (G.38)$$

Chapter 14

1. In the numerator we use $dx\,dy\,dz\,dp_x\,dp_y\,dp_z = dr\,d\theta\,d\phi\,dp_r\,dp_\theta\,dp_\phi$. Thus the integral

$$\int d\mathbf{r} \int d\mathbf{p}_r \exp(-\beta H)\delta(r - r^*)\left(-\frac{p_r}{m}\right)\theta\left(-\frac{p_r}{m}\right)$$
$$= \exp\{-\beta[U_{AB}(r^*) + V_{int} + U]\}(2\pi)^2 m(kT)^2 2r^{*2}$$
$$= \exp[-\beta(U_{AB} + V_{int} + U)]kTh^2 Q_{TS,\,rot} \qquad (G.39)$$

where

$$Q_{TS,\,rot} = \frac{2I^* kT}{\hbar^2} \qquad (G.40)$$

and I^* the moment of inertia at the transition state. In the denominator we can integrate over $d\mathbf{p}_r$, i.e.

$$\int dp_x \int dp_y \int dp_z \exp\left(-\frac{\beta}{2m}(p_x^2 + p_y^2 + p_z^2)\right) = (2mkT)^{3/2} = h^3 Q_{R,\,\text{rel}} \quad \text{(G.41)}$$

The momentum space for the solvent can be integrated over in the same manner in both numerator and denominator. These two factors cancel each other.

2. The Laplacian is, in polar coordinates,

$$\frac{1}{r}\frac{\partial^2}{\partial r^2}r + \frac{1}{r^2 \sin\theta}\frac{\partial}{\partial\theta}\left(\sin\theta\frac{\partial}{\partial\theta} + \frac{1}{r^2 \sin^2\theta}\frac{\partial^2}{\partial\phi^2}\right) \quad \text{(G.42)}$$

For spherically symmetric problems the last two operators can be omitted. Hence the diffusion equation is

$$\frac{\partial\rho}{\partial t} = \frac{D}{r}\frac{\partial^2}{\partial r^2}(r\rho) \quad \text{(G.43)}$$

and the Laplace transform is

$$\underline{\rho} = \frac{1}{s} + \frac{B(s)}{r}\exp(-r\sqrt{s/D}) \quad \text{(G.44)}$$

The Laplace transform at $r = R$ gives

$$B = -\frac{R}{s}\exp(R\sqrt{s/D}) \quad \text{(G.45)}$$

from which the solution can be obtained by inverse transformation.

3. $k_d = 4\pi DR = 5.0 \cdot 10^{-12} \text{ cm}^3/\text{sec}.$

The term $R/\sqrt{\pi Dt}$ can be used to estimate the time scale, which is on the order of nanoseconds.

$$\frac{1}{k} = \frac{1}{k_{\text{act}}} + \frac{1}{4\pi RD}$$

yields $k_d = 0.33 \cdot 10^{-11} \text{ cm}^3/\text{sec}.$

4. Use $D = kT/6\pi\eta a$ to obtain $k_d = 8kT/3\eta$. Thus we get $k_d = 0.46 \cdot 10^{-10}$ cm^3/sec, $9.0 \cdot 10^{-12}$ cm^3/sec, and $7.2 \cdot 10^{-15}$ cm^3/sec at 300 K for ether, ethanol, and glycerol respectively.

5. a. $k = 4\pi RD = 1.38 \cdot 10^{-10}$ cm^3/sec

    b. $C_A(R) = 0$ and $C_A(\infty) = C_A^*$. $I(r) = $ constant if $\delta = z_A z_B e^2/\epsilon k_B T$.

    c. $\delta \rightarrow 0$ $C_A(r) = C_A^*(1 - R/r)$, $I(R) = C_A^* 4\pi D_A R$.

    d. $k(T) = 4\pi D_A R\Delta/[\exp(\Delta) - 1]$; $\Delta = \delta/R$. At $k = 298$ K, $k = 2.12 \cdot 10^{-10}$ cm^3/sec.

6. a.
$$\frac{\partial c_p}{\partial t} = D \frac{\partial^2 c_p}{\partial x^2} \tag{G.46}$$

    with the boundary conditions

$$t = 0, x \geq 0: \quad c_p = c^* \tag{G.47}$$

$$t \geq 0, x \rightarrow \infty: \quad c_p \rightarrow c^* \tag{G.48}$$

$$t > 0, x = 0: \quad D \frac{\partial c_p}{\partial x} = k_{ET} c_p(x = 0) \tag{G.49}$$

    The current is given by

$$i = FAk_f c_p(x = 0) \tag{G.50}$$

    b.
$$\left(\frac{\partial \overline{c_p}}{\partial x} \right)_{x=0} = -\sqrt{\frac{S}{D}} \left[\overline{c_p}(x = 0) - \frac{c^*}{S} \right] \tag{G.51}$$

    c.
$$\overline{c_p}(x = 0) = c^* \exp\left(\frac{k_{ET}^2}{D} t \right) \cdot \operatorname{erfc}\left(\frac{k_{ET}}{\sqrt{D}} \sqrt{t} \right) \tag{G.52}$$

Chapter 15

1. Consider a particle in equilibrium with the surroundings. We then have

$$M \frac{dv}{dt} = F(t) \tag{G.53}$$

or

$$M\Delta v = \int_t^{t+\tau} F(t') \, dt' \tag{G.54}$$

from which we get

$$M^2\langle\Delta v^2\rangle = \left\langle \int_t^{t+\tau} dt'\, F(t') \int_t^{t+\tau} dt''\, F(t'') \right\rangle \tag{G.55}$$

Introducing $t'' = t' + s$ we have

$$\int_t^{t+\tau} dt' \int_{t-t'}^{t-t'+\tau} ds \langle F(t')F(t'+s)\rangle \tag{G.56}$$

which can be written as

$$\tau \int_{-\infty}^{\infty} \langle F(0)F(s)\rangle ds = \tau 2kT\gamma M \tag{G.57}$$

where we have used Eq. (15.5).

2. Use $Y = Y[v - a(x - x_b)]$ and obtain from Eq. (15.36)

$$Y'[v(a-\gamma) - \omega_b^2(x-x_b)] + \frac{\gamma kT}{M} Y'' = 0 \tag{G.58}$$

where $Y' = \partial Y/\partial\tilde{u}$ and $Y'' = \partial^2 Y/\partial\tilde{u}^2$. Introducing $v(a-\gamma) - \omega_b^2(x-x_b) = (a-\gamma)\tilde{u}$ we get the solution, Eq. (15.37). If we insert $\tilde{u} = v - a(x-x_b)$ in this equation we obtain:

$$a = \frac{\gamma}{2} \pm \sqrt{\frac{\gamma^2}{4} + \omega_b^2} \tag{G.59}$$

where the lower sign must be discarded since $a > 0$. Note that

$$\tilde{u}(x_c) = v - a(x_c - x_b) \sim -\infty \tag{G.60}$$

and

$$\tilde{u}(x_a) = v - a(x_a - x_b) \sim \infty \tag{G.61}$$

Thus

$$Y(x_a) \sim K \int_{-\infty}^{\infty} d\tilde{u}\exp\left(-\frac{\lambda_+ m\beta\tilde{u}^2}{2\gamma}\right) = 1 \tag{G.62}$$

if $K = \sqrt{m\beta\lambda_+/2\pi\gamma}$

3. Using

$$n_a = \int_{-\infty}^{\infty} dx\, dv\, P(x,v) \tag{G.63}$$

the expression for $P(x,v)$, and $u(x) = u(x_a) + \frac{1}{2}M\omega_0^2(x - x_a)^2$, we get expression (15.39).

4. Using

$$j = \int_{-\infty}^{\infty} dv\, vP(x_b,v) \tag{G.64}$$

and

$$P(x_b,v) = \frac{1}{Q} Y(x_b,v)\exp\left\{-\beta\left[\frac{M}{2}v^2 + U(x_b)\right]\right\} \tag{G.65}$$

we get

$$j = K\int_{-\infty}^{\infty} dv\, v\, \frac{1}{Q}\exp\left(-\beta Mv^2/2\right)\int_{-\infty}^{v} du$$
$$\cdot \exp\left(-\lambda_+Mu^2/2\gamma kT\right)\exp\left[-\beta U(x_b)\right] \tag{G.66}$$

which gives

$$j = \frac{K}{Q}\sqrt{2\pi}\,\frac{kT}{M}\exp\left[-\beta U(x_b)\right]\sqrt{\frac{\gamma kT}{Ma}} \tag{G.67}$$

where we have used that

$$\int_{-\infty}^{\infty} dv\, v\exp\left(-Av^2\right)\int_{-\infty}^{v}\exp\left(-Bu^2\right)du$$
$$= 2\int_{0}^{\infty} dv\, v\exp\left(-Av^2\right)\int_{0}^{v}\exp\left(-Bu^2\right)du \tag{G.68}$$

5. We note that $dU/dx = 0$ for $x = 0$ and $x = \pm\sqrt{a_0/c_0}$. From the second derivative at $x = 0$ we have $\omega_b^2 = a_0/M = 415$ kJ mol^{-1} amu^{-1} Å^{-2} = 1081 cm^{-1}. The barrier height is obtained as

$$E_b = U(x = 0) - U(x = \sqrt{a_0/c_0}) = 26.3 \, \text{kJ/mol} \qquad (\text{G.69})$$

The frequency at the minimum is $\omega_0 = \sqrt{2a_0} = 1529 \, \text{cm}^{-1}$. The transition state rate constant obtained is

$$k^{TST} = \frac{\omega_0}{2\pi} \exp(-\beta E_b) = 1.21 \cdot 10^9 \, \text{sec}^{-1} \qquad (\text{G.70})$$

and the Kramers correction factor:

$$k^{Kr} = \frac{1}{\omega_b} \left\{ \sqrt{\omega_b^2 + \frac{\gamma^2}{4}} - \frac{\gamma}{2} \right\} \qquad (\text{G.71})$$

Using $D = kT/M\gamma = 10^{-4} \, \text{cm}^2/\text{sec}$, we obtain $k^{Kr} = 0.560$.

6. The expression is obtained in the high-temperature or low-frequency limit of vibrational partition functions. The assumption is that the frequencies of the system (solvent + reactants) are unchanged at the transition state. $\kappa \rightarrow \omega_b/\gamma$ (low friction); $\kappa \rightarrow (\omega_b/\gamma)^2$ (high friction).

7. $L = 3.19 \, \text{Å}$, $\gamma = 1.73 \cdot 10^{14} \, \text{sec}^{-1}$, and $D = 1.43 \cdot 10^{-4} \, \text{cm}^2/\text{sec}$.

Chapter 16

1. Use the steady state for the concentrations of $[A^+|B^-]^*$ and $[A|B]^*$.

2. $a_1 = a_2 = a$ gives $E_A = U_2(R^*) - U_1^0 = (1/4\lambda)(\Delta U^0 + \lambda)^2$, where $\Delta U_0 = U_2^0 - U_1^0$ and $\lambda = a(R_1^0 - R_2^0)^2$. R^* is defined by $U_1(R^*) = U_2(R^*)$.

4. **a.** $\Delta F^0 = 0$ and

$$k_{ET} = Z \exp\left[-\frac{1}{8a} \left(\frac{1}{\epsilon_{op}} - \frac{1}{\epsilon_{st}} \right) e^2/kT \right] \qquad (\text{G.72})$$

where $a = a_i$. For $T = 300 \, \text{K}$, we have $k_{ET} = 4.48 \cdot 10^5 \, \text{sec}^{-1}$ for $a = 2$ Å, $1.06 \cdot 10^8 \, \text{sec}^{-1}$ for $a = 3$ Å, and $1.64 \cdot 10^9 \, \text{sec}^{-1}$ for $a = 4$ Å.

b. $\lambda_{AD} = (\lambda_A^2 + \lambda_D^2)/(\lambda_A + \lambda_D)$.

5. **a.**
$$E_{in}^* = \frac{6}{2} f_2(d_2^* - d_2^o)^2 + \frac{6}{2} f_3(d_3^* - d_3^o)^2 \qquad (\text{G.73})$$

and energy conservation gives $d_2^* = d_3^* = d^*$.

b. Minimizing the potential energy gives

$$d^* = \frac{f_2 d_2^o + f_3 d_3^o}{f_2 + f_3} \tag{G.74}$$

and the inner-shell reorganization energy is given by

$$E_{in}^* = \frac{3 f_2 f_3 (\Delta d^o)^2}{f_2 + f_3} \tag{G.75}$$

where $\Delta d^o = d_2^o - d_3^o$.

c. The inner-shell reorganization energy for a pair of compounds with n identical ligands is

$$E_{in}^* = \frac{n f_2 f_3 (\Delta d^o)^2}{2(f_2 + f_3)} \tag{G.76}$$

6. a.
$$k_{diff}(t) = 4\pi D(a_M + a_N)\left(1 + \frac{a_M + a_N}{\sqrt{\pi D t}}\right) \tag{G.77}$$

$$k_{diff} = 4\pi D(a_M + a_N) \tag{G.78}$$

b.
$$k_{act} = \frac{k \, k_{diff}}{k_{diff} - k} \tag{G.79}$$

c.
$$E_{pol} = -\frac{1}{2} Z_Q^2 \frac{1}{a_Q}\left(1 - \frac{1}{\epsilon_{st}}\right) \tag{G.80}$$

for $Q = M, N$.

d.
$$\lambda = \left(\frac{1}{2a_M} + \frac{1}{2a_N} - \frac{1}{a_M + a_N}\right)\left(\frac{1}{\epsilon_{op}} - \frac{1}{\epsilon_{st}}\right) \tag{G.81}$$

λ is calculated for a solvent that is in a nonequilibrium configuration, and the dielectric solvation energies are calculated for the solvent in an equilibrium configuration.

$$\frac{(\Delta F^o + \lambda)^2}{4\lambda} + \frac{e^2 Z_M Z_N}{R \epsilon_{st}} \tag{G.82}$$

e. $k_{ET} = \dfrac{k_B T}{h} \exp\left[-\dfrac{1}{4kT}\left(\dfrac{1}{2a_M} + \dfrac{1}{2a_N} - \dfrac{1}{R}\right)\left(\dfrac{1}{\epsilon_{op}} - \dfrac{1}{\epsilon_{st}}\right)\right]$

$$\cdot \exp\left(-\frac{e^2 Z_M Z_N}{R \epsilon_{st} k_B T}\right) \tag{G.83}$$

The electronic factor and its R-dependence have been neglected.

INDEX